SpringerBriefs in Geography

SpringerBriefs in Geography presents concise summaries of cutting-edge research and practical applications across the fields of physical, environmental and human geography. It publishes compact refereed monographs under the editorial supervision of an international advisory board with the aim to publish 8 to 12 weeks after acceptance. Volumes are compact, 50 to 125 pages, with a clear focus. The series covers a range of content from professional to academic such as: timely reports of state-of-the art analytical techniques, bridges between new research results, snapshots of hot and/or emerging topics, elaborated thesis, literature reviews, and in-depth case studies.

The scope of the series spans the entire field of geography, with a view to significantly advance research. The character of the series is international and multidisciplinary and will include research areas such as: GIS/cartography, remote sensing, geographical education, geospatial analysis, techniques and modeling, landscape/regional and urban planning, economic geography, housing and the built environment, and quantitative geography. Volumes in this series may analyze past, present and/or future trends, as well as their determinants and consequences. Both solicited and unsolicited manuscripts are considered for publication in this series.

SpringerBriefs in Geography will be of interest to a wide range of individuals with interests in physical, environmental and human geography as well as for researchers from allied disciplines.

More information about this series at https://link.springer.com/bookseries/10050

Anita De Franco

Abandonment as a Social Fact

The Problem of Unused and
Unmaintained Private Buildings
in a Neo-institutional Perspective

 Springer

Anita De Franco ⓘ
Department of Architecture and Urban
Studies (DASTU)
Polytechnic University of Milan
Milan, Italy

ISSN 2211-4165 ISSN 2211-4173 (electronic)
SpringerBriefs in Geography
ISBN 978-3-030-90366-4 ISBN 978-3-030-90367-1 (eBook)
https://doi.org/10.1007/978-3-030-90367-1

This Springer imprint is published by the registered company Springer Nature Switzerland AG
The registered company address is: Gewerbestrasse 11, 6330 Cham, Switzerland

Contents

Chapter 1
Introduction

Abstract This introduction is divided into five sections. The first sets the focus: abandonment as a social fact (Sect. 1.1). The second explains the approach: a neo-institutional perspective on urban issues (Sect. 1.2). The third exposes the research question: what changes once we recognise abandonment as a social fact? (Sect. 1.3). The fourth is devoted to the methodology (Sect. 1.4). The fifth presents the structure of the book (Sect. 1.5).

Keywords Social facts · Institutions · Institutional frameworks · Neo-institutionalism

1.1 Focus

As is well known, the problem of the abandonment of buildings is much discussed in the literature and public debate. However, the impression is that there is a certain imprecision in this regard, for example in the use of certain terms and concepts—e.g. abandoned, vacant, empty, derelict, deserted, decommissioned buildings—that are not always unequivocal.[1]

Some alarming estimates of the phenomena considered here do not distinguish sharply among different cases (which are—for instance in Italy and particularly in urban situations—on totally different magnitudes). Certain debates tend to equate vacant/abandoned buildings with those buildings "at their owners' disposal but not used on a continuous basis" (Agenzia delle Entrate 2017a). In Italy, this category of buildings accounted for over 20% of the properties in 3630 municipalities in 2014. The main issue, however, is that this category is an even broader vague one, employed to accentuate the situation dramatically. It includes either merely empty

---

[1] As also stressed by Morckel (2014), Hwang and Lee (2019), Moroni et al. (2020a, b), Buitelaar et al. (2021), Chiodelli and Caramaschi (2021). Hwang and Lee (2019, p. 2) write: "Since they lack standard definitions, defining the objects and scopes of urban voids and other similar terms [is] difficult .... Multiple discourses have extensively used many interchangeable terms to denote urban voids ... Therefore, as some scholars have insisted, there is an urgent need for critical analysis, appropriate approaches and further insightful systematic analysis regarding urban voids".

and truly abandoned buildings, but also those that are used occasionally, e.g. holiday homes (Moroni et al. 2020b). Another example here is the Eurostat report (2015). This report underscores emphatically that almost one in six dwellings in Europe are unoccupied; but, according to it, dwellings are counted as unoccupied "if they are reserved for seasonal or secondary use (such as holiday homes) or if they are vacant (dwellings which may be for sale, for rent, for demolition, or simply lying empty and unused)" (Eurostat 2015, p. 75).[2]

This conceptual confusion does not help to understand the nature of the problem, to estimate the size of the problem and to address it (when and how to intervene). A reconceptualization of the very idea of "abandonment" therefore seems necessary.

In revisiting the issue this study will focus on: (i) *private properties* that are owned by single or multiple social entities (e.g. individuals, firms, enterprises); (ii) *buildings* (i.e. fully built structures); (iii) *urban contexts* (i.e. dense, compact, socio-economically stable settings).

As a consequence: publicly owned abandoned buildings are a priori excluded[3]; halted sites, empty lots or land parcels are not considered here[4]; the contexts under examination can be assumed as "non-shrinking" cities where abandonment, differently from "shrinking" cities, does not necessarily result from large-scale depopulation patterns (e.g. migration).

1.2 Approach

This study proposes to define as "abandoned" a building whose *owner* does not fulfil his/her *responsibilities* ensuing from ownership. In this perspective, abandonment is considered a *social fact* (Searle 1995).

The work adopts—and tries to further develop—a *neo-institutional* view on the phenomenon.[5] Actually, neo-institutionalism is reinterpreted here as an attempt *to take certain social facts seriously.*

---

[2] Clearly, in this case as in others, each and every estimate depends on how the phenomenon considered is defined Morckel (2014).

[3] These also are a burdening problem in Italy (see, e.g. Moroni et al. 2020b).

[4] These also are a crucial and interesting problem. See Greenberg et al. (1990), Bowman and Pagano (2000), Ho and Spoor (2006), Durst and Ward (2015), Humphris and Rauws (2020). Moreover, the focus on "buildings" refers to any type of property independently from its functional classification. Previous studies instead have adopted a more sectorial approach focused on housing/residential structures (Edson 1972; Lieb et al. 1974; Morgan 1980; White 1986; Wilson et al. 1994; Cohen 2001; Hillier et al. 2003; vom Hofe et al. 2019) and/or types of non-residential structures (e.g. brownfields: Németh and Langhorst 2014; Haase et al. 2014, p. 1524; Dixon and Adams 2008).

[5] As well-known, the term "neo-institutionalism" is usually employed to distinguish the recent renewed interest in institutions from the so-called "old-institutionalism". Of especial interest in general are here the neo-institutional theories developed by North (1990) and Kasper and Streit (1998). With particular reference to planning, see Alexander (2001, 2002, 2005), Pennington (2003),

Part of the discussion in the field of planning and urban policies began to explore the issue of abandonment paying attention to institutional issues.[6] The attempt of this book is to go on critically in this direction. The phenomenon is questioned considering the "rules of the game" at stake and the functional expectations assigned to objects and people (Searle 1995, pp. 27 ff.). Consider the following two statements: (i) abandonment signals the inadequate working of social entities (e.g. private owners, properties); and (ii) abandonment signals the inadequate working of social activities (e.g. urban systems, real estate markets). Both statements may point to the same problem, but we often run the risk of overlapping them. On this, see Moroni (2018, 2019).

Three features of the neo-institutional approach here applied to urban issues are the following.

Firstly, to acknowledge, in a more explicit way, the philosophical dimension of certain issues. In this regard, the impression is that John Searle's work has much to teach to planning theories and urban studies.[7] The assumption is that certain urban aspects—such as the abandonment of buildings—directly depend on the way social agents react to (interpret) the context they inhabit, including the normative one.[8]

Secondly, to place the relationship between objects and people at the centre of attention, re-including objects in the analysis (thus, accepting Latour's 2005 invitation,[9] albeit in an original way). Objects are not considered as mere "participants" to social action, as they are invested with special deontic qualifications.[10]

Buitelaar (2003, 2007), Lai (2005), Moroni (2010, 2018, 2019, 2020), Dembski and Salet (2010), Sorensen (2015), Savini et al. (2015), Salet (2018a, b), Moroni and Minola (2019).

[6] In discussing the issue of abandonment, the work of the following authors is based on issues like property rights, liability and tort rules: Lieb et al. (1974), Sternlieb et al. (1974), Stone (1985), Heller (1998), Kraut (1999). Other authors, subsequently, highlighted the problem in term of taxation, credit, debts, loans, foreclosures, with particular reference to post-crisis shrinking situations: Basset et al. (2006), Immergluck and Smith (2006), Samsa (2008), Schilling (2009), Mallach (2012), Silverman et al. (2013), Dewar and Weber (2012), Aiyer et al. (2015), Grossmann and Haase (2016), Wang and Immergluck (2019). Recently, some authors have been more focused on local issues of organization and rules trying to cope with the problem: Sampson et al. (2017), Gentili and Hoekstra (2018), Micelli and Pellegrini (2017, 2019), Foster and Newell (2019), Galster (2019), Hwang and Lee (2019), Jeon and Kim (2020).

[7] Effectively, certain insights of his work have not been taken into consideration in urban studies (among the few exceptions see Madanipour 1999; Smith 2020a, b). To a certain extent, one may describe a large part of urban elements in terms of social/institutional facts (administrative boundaries, governments, organizations, associations, properties, contracts, laws, rules, conventions, traditions, habits, etc.).

[8] The overall idea is that abandonment is a man-made problem and as such should be treated. Compare with Popper (1974, pp. 208 ff.).

[9] In urban studies, see especially Lieto (2017) and Beauregard (2012, 2015).

[10] The approach adopted here recognizes the primacy of social processes over material items. In this work the analysis of abandonment does not therefore suggest any parallelism between physical and social issues in cities. (While this may seem obvious, scholarly debates do not always recognize how stressing the difference between *physical* and *social* dimension is helpful to reorganize knowledge and research: Smith 2020b, p. 20).

Thirdly, to accept, in a more radical way, the dynamic aspect of urban phenomena, trying to combine the attention to institutions with theories of urban complexity (along the line proposed by Moroni 2015a).[11] Considering abandonment as an extraordinary and intrinsically undesirable situation, authorities feel often compelled to reconstitute an optimal state of affairs.[12] In contrast, the approach used here suggests a more holistic, dynamic understanding of abandonment as part of a lively world.

1.3   Research Questions

The central question of this study is: What changes in the interpretation of and intervention on, the problem of abandonment once we recognize it as a *social fact*?

From a neo-institutional perspective, the general research question can be articulated in a series of sub-questions, such as:

1. to what extent can the institutional framework be co-responsible in the emergence of the phenomenon of abandonment and
2. what modifications to the institutional framework may favour the recovery of what now lies abandoned?

In this work, we refer to *institutional frameworks* as sets of general rules of conduct that coordinate the actions of social agents (in urban studies, see, e.g. Moroni 2010, Salet 2018a).[13]

---

[11] See also Moroni and Cozzolino (2020), Cozzolino and Moroni (2021), Rauws et al. (2020) and Zhang et al. (2020).

[12] Without recognizing the social nature of the abandonment problem, one may jump to the conclusion that the issue must be solved for the sake of the "efficiency" of urban processes. The use of "efficiency" as a normative value to drive public decisions or planning endeavours is seen by some as a useful criterion (Lai 2005). However, for other scholars, our preference for an efficient (material) world inevitably clashes with the complexity of the social world (De Roo and Rauws 2012).

[13] The recurrent ignorance around "institutions" can be explained by the fact that these are not self-evident: they require *recognition* (they pertain to an "invisible ontology": Searle 1995, 2010). On the non-immediate visibility of constitutive rules, see also Conte (2011).

1.4   Methodology

This study is mainly analytical and conceptual, yet it is empirically informed and illustrated. From a methodological point of view, the work is based on: an extensive literature review; reports and documents of Italian[14] and international[15] public and private institutions and agencies; an analysis of Italian laws and regulations; interviews with key actors[16]; one main case study: the case of Milan.[17]

---

[14] For instance, Agenzia delle Entrate (2015, 2017a, b, 2019a, b), Banca d'Italia (2019), ANCE (2012, 2019), Cittalia and ANCI (2009), Confedilizia (2015, 2019), Legambiente and CRESME (2013), ISPRA (2015), ISTAT (2014), Ministry of Economics and Finances (MEF 2017 2018, 2019), WWF Italy (2013). These reports and documents have been chosen because they are the main sources of information on the problem in question (in Italy). Some of them are produced by public agencies that deal with data collection and publications on socioeconomic and geographic data (i.e. ISTAT, ISPRA); other public agencies deal more directly with tax and real estate matters (i.e. Agenzia delle Entrate, Banca d'Italia, MEF); other reports derive from surveys of private agencies and trade associations (i.e. ANCE, Confedilizia), and other organizations of general interest (i.e. Cittalia and ANCI, Legambiente and CRESME, WWF Italy).

[15] For instance Buildings Performance Institute Europe (BPIE 2011), Eurostat (2015, 2018), PWC (2020), World Bank (2019). These reports and documents have been chosen because they were produced by international reliable agencies.

[16] The ten semi-structured interviews regarding the Italian situation, and Milan in particular, lasted around 30 min with each interviewee. In this regard, I am particularly grateful to Pierfrancesco Maran (Assessor of Urban Planning, Green Areas and Agriculture, Municipality of Milan), Giorgio Spaziani Testa (President of *Confedilizia*, i.e. the Italian Confederation of Property Owners), Marcello Cruciani (*ANCE*, i.e. the National Association of Italian Construction Contractors), Filomena Pomilio (architect and adviser of the *Ordine degli architetti*, Milan), Uberto Visconti (real estate developer), Luca Minola (investment analyst, Round Hill Capital), Marzia Morena (professor and president of *Federimmobiliare* from 2014 to 2016), Emanuele Garda (*Laboratorio permanente sui luoghi dell'abbandono*: *L'ABB*, Milan), Matteo Goldstein Bolocan and Franco Sacchi (president and director of *Centro Studi PIM*, Milan). The interviewees have been chosen on the basis of their skills and roles—crucial for the subject matter—paying particular attention to the diversity of points of view (e.g. owners, developers, professionals, public officials, politicians). Additional interviews were held with specialized technicians and other actors to compare the situation in different countries, e.g. the Netherlands. In this regard, I am particularly grateful to Hilde Remøy, Tom Daamen (TU Delft) and Sanne Holtslag-Broekhof (*Kadaster*). Obviously, the views expressed in this work are solely those of the author. The usual disclaimers apply.

[17] Reconstructed through an overview of the local sections of national newspapers in the past 5 years (especially, *Corriere* and *Repubblica*), reports of local agencies and organizations (e.g. Confcommercio 2015, Camera di Commercio 2017, 2018, 2019), an analysis of local normative records (e.g. Comune di Milano 2016, 2017, 2018a, b, 2019a, b, c), GIS mappings, photographs, satellite imagines. On-the-spot observations were also carried with specific attention to the city of Milan (especially during October 2017–2018) and on other contexts abroad (e.g. Amsterdam, during the visiting periods at the University of Utrecht in the Netherlands, February-May 2019).

1.5 Structure

The book is structured in the following way: Chap. 2 presents the theoretical framework and Chap. 3 an analytical schema; Chap. 4 outlines the main findings and implications and Chap. 5 concludes. Two appendixes present additional materials on the general situation in Italy and the case study of Milan, in the Lombardy region (Appendix A and B).[18]

References

Agenzia delle Entrate (2015) La tassazione immobiliare: un confronto internazionale [Real estate taxation: an international comparison]. Available at: https://www.finanze.gov. Accessed 13 Oct 2018

Agenzia delle Entrate (2017a) Gli immobili in Italia [Real estate in Italy]. Available at: http://www.mef.gov.it. Accessed 24 Maggio 2019

Agenzia delle Entrate (2017b) Statistiche catastali 2016 [Cadastral statistics 2016]. Available at: https://wwwt.agenziaentrate.gov.it. Accessed 2 June 2019

Agenzia delle Entrate (2019a) Rapporto immobiliare 2019: settore residenziale [Real estate report 2019. Residential sector]. Available at: https://www.agenziaentrate.gov.it. Accessed 27 Jan 2020

Agenzia delle Entrate (2019b) Statistiche catastali 2018: catasto edilizio urbano [Land registry report 2019]. Available at: www.agenziaentrate.gov.it. Accessed 27 Jan 2020

Aiyer SM, Zimmerman MA, Morrel-Samuels S, Reischl TM (2015) From broken windows to busy streets: a community empowerment perspective. Health Educ Behav 42(2):137–147. https://doi.org/10.1177/1090198114558590

Alexander ER (2001) Why planning versus markets is an oxymoron: asking the right question. Plann Mark 4(1):1–8

Alexander ER (2002) Planning rights: towards normative criteria for evaluating plans. Int Plan Stud 7(3):191–212

Alexander ER (2005) Institutional transformation and planning: from institutionalization theory to institutional design. Plan Pract Res 4(3):209–223. https://doi.org/10.1177/1473095205058494

ANCE (2012) Osservatorio congiunturale sull'industria delle costruzioni [Economic observatory on the construction industry]. Available at: http://www.ance.it. Accessed 28 Oct 2018

ANCE (2019) Osservatorio congiunturale sull'industria delle costruzioni [Economic observatory on the construction industry]. Available at: https://www.ance.it. Accessed 15 Nov 2019

Banca d'Italia (2019) Finanza pubblica: Fabbisogno e debito [Public finance: needs and debt]. Available at: https://www.bancaditalia.it. Accessed 5 March 2020

Basset EM, Schweitzer J, Panken S (2006) Understanding housing abandonment and owner decision-making in Flint, Michigan: an exploratory analysis. Lincoln Inst Land Policy (sept). https://doi.org/10.1017/CBO9781107415324.004

Beauregard RA (2012) Planning with things. J Plan Educ Res 32(2):182–190

Beauregard RA (2015) Planning matter: acting with things. University of Chicago Press, Chicago

Bowman AOM, Pagano MA (2000) Transforming America's cities: policies and conditions of vacant land. Urban Affairs Rev 35(4):559–581

BPIE (Buildings Performance Institute Europe) (2011) Europe's buildings under the microscope. A country-by-country review of the energy performance of buildings. Available at: http://bpie.eu. Accessed 18 Sept 2019

---

[18] These appendixes are partially based on ideas and materials developed by the author in Moroni et al. (2020b) and De Franco (2021).

Buitelaar E (2003) Neither market nor government: comparing the performance of user rights regimes. Town Plann Rev 74(3):315–330

Buitelaar E (2007) The cost of land use decisions: applying transaction cost economics to planning and development. Blackwell Publishing, Oxford

Buitelaar E, Galle M, Sorel N (2011) Plan-led planning systems in development-led practices: an empirical analysis into the (lack of) institutionalisation of planning law. Environ Plan A 43(4):928–941

Buitelaar E, Moroni S, De Franco A (2021) Building obsolescence in the evolving city. Reframing property vacancy and abandonment in the light of urban dynamics and complexity. Cities 108 (online ahead of print)

Camera di Commercio (2017) Milano, l'Europa. Città internazionali a confronto [Milan, Europe. International cities in comparison]. Available at: https://www.milomb.camcom.it. Accessed 19 June 2020

Camera di Commercio (2018) L'importanza di essere una start up [The importance of being a start-up]. Available at: https://www.milomb.camcom.it. Accessed 19 June 2020

Camera di Commercio (2019) Milano produttiva [Productive Milan]. Available at: https://www.milomb.camcom.it. Accessed 19 June 2020

Chiodelli F, Caramaschi S (2021) Housing emptiness beyond vacancy and abandonment: conceptual and definitional points in question (forthcoming)

Cittalia, ANCI (2009) Oltre le ordinanze i sindaci e la sicurezza urbana. Cittalia Fondazione Anci ricerche. Available at: https://www.cittalia.it. Accessed 14 Jan 2018

Cohen JR (2001) Abandoned housing: exploring lessons from Baltimore. Hous Policy Debate 12(3):415–448. https://doi.org/10.1080/10511482.2001.9521413

Comune di Milano (2016) Regolamento edilizio [Building regulations]. Available at: http://www.comune.milano.it. Accessed 27 March 2020

Comune di Milano (2017) Documento di visione strategica: scali ferroviari [Strategic vision document. Railway yards]. Available at: http://www.comune.milano.it. Accessed 21 March 2020

Comune di Milano (2018a) Documento unico di programmazione 2019–2012 [Single programming document 2019–2012]. Available at: http://www.comune.milano.it. Accessed 4 April 2019

Comune di Milano (2018b) Ricognizione sullo stato di attuazione dei programmi [Reconnaissance on the state of implementation of the programs]. Available at: http://www.comune.milano.it. Accessed 6 May 2019

Comune di Milano (2019a) Documento di piano [Plan document]. Available at: http://www.comune.milano.it. Accessed 18 July 2020

Comune di Milano (2019b) Piano delle regole [Rules plan]. Available at: http://www.comune.milano.it. Accessed 18 July 2020

Comune di Milano (2019c) Piano dei servizi [Service plan]. Available at: http://www.comune.milano.it. Accessed 18 July 2020

Confcommercio (2015) Report attività per Expo Milano 2015 [Report for Expo Milano 2015]. Available at: https://www.confcommerciomilano.it. Accessed 30 Sept 2020

Confedilizia (2019) Confedilizia notizie [multiple issues]

Confedilizia (2015) Dossier tassazione immobili [Property tax dossier]. Available at: https://www.confedilizia.it. Accessed 16 May 2018

Conte AG (2011) Sociologia filosofica del diritto. Giappichelli, Torino

Cozzolino S, Moroni S (2021) Multiple agents and self-organisation in complex cities: the crucial role of several property. Land Use Policy 103:1–7

De Franco A (2021) Addressing the problem of private abandoned buildings in Italy. A neo-institutional approach to multiple causes and potential solutions. In: Bisello A, Vettorato D, Haarstad H, Borsboom-van Beurden J (eds) Smart and sustainable planning for cities and regions: results of SSPCR 2019, vol 2. Springer, Berlin, pp 235–247

de Roo G, Rauws WS (2012) Positioning planning in the world of order, chaos and complexity: on perspectives, behaviour and interventions in a non-linear environment. In: Portugali J, Meyer H,

Stolk E, Tan E (eds) Complexity theories of cities have come of age: an overview with implications to urban planning and design. Springer, Heidelberg, pp 207–220

Dembski S, Salet W (2010) The transformative potential of institutions: how symbolic markers can institute new social meaning in changing cities. Environ Plan A 42(3):611–625. https://doi.org/10.1068/a42184

Dewar M, Weber MD (2012) City abandonment. In: Dewar M, Weber MD (eds) The Oxford handbook of urban planning. Oxford University Press, Oxford, pp 563–586

Dixon T, Adams D (2008) Housing supply and brownfield regeneration in a post-Barker world: is there enough brownfield land in England and Scotland? Urban Studies 45(1):115–139

Durst NJ, Ward PM (2015) Lot vacancy and property abandonment: colonias and informal subdivisions in Texas. Int J Hous Policy 15(4):377–399. https://doi.org/10.1080/14616718.2015.1090095

Edson CL (1972) Housing abandonment—the problem and a proposed solution. Real Property, Probate Trust J 7(2):382–390

Eurostat (2015) People in the EU: who are we and how do we live? European Union. Available at: https://ec.europa.eu/eurostat. Accessed 25 May 2020

Eurostat (2018) Living conditions in Europe. Statistical books. Publications Office of the European Union, Luxembourg

Foster A, Newell JP (2019) Detroit's lines of desire: footpaths and vacant land in the motor city. Landsc Urban Plan 189(April):260–273. https://doi.org/10.1016/j.landurbplan.2019.04.009

Galster G (2019) Why shrinking cities are not mirror images of growing cities: a research agenda of six testable propositions. Urban Affairs Rev 55(1):355–372. https://doi.org/10.1177/1078087417720543

Gentili M, Hoekstra J (2018) Houses without people and people without houses: a cultural and institutional exploration of an Italian paradox. Hous Stud 34(3):425–447. https://doi.org/10.1080/02673037.2018.1447093

Greenberg MR, Popper FJ, West BM (1990) The TOADS a new American urban epidemic. Urban Affairs Q 25(3):435–454. https://doi.org/10.1177/004208169002500306

Grossmann K, Haase A (2016) Neighborhood change beyond clear storylines: what can assemblage and complexity theories contribute to understandings of seemingly paradoxical neighborhood development? Urban Geogr 37(5):727–747. https://doi.org/10.1080/02723638.2015.1113807

Haase A, Rink D, Grossmann K, Bernt M, Mykhnenko V (2014) Conceptualizing urban shrinkage. Environ Plan A 46:1519–1534. https://doi.org/10.1068/a46269

Heller MA (1998) The tragedy of the anticommons: property in the transition. Harv Law Rev 111(3):621–688

Hillier AE, Culhane DP, Smith TE, Tomlin CD (2003) Predicting housing abandonment with the Philadelphia neighborhood information system. J Urban Aff 25(1):91–106. https://doi.org/10.1111/1467-9906.00007

Ho P, Spoor M (2006) Whose land? The political economy of land titling in transitional economies. Land Use Policy 23(4):580–587. https://doi.org/10.1016/j.landusepol.2005.05.007

Humphris I, Rauws W (2020) Edgelands of practice: post-industrial landscapes and the conditions of informal spatial appropriation. Landscape Res 46(5):589–604. https://doi.org/10.1080/01426397.2020.1850663

Hwang SW, Lee SJ (2019) Unused, underused, and misused: an examination of theories on urban void spaces. Urban Res Pract 13(5):1–17. https://doi.org/10.1080/17535069.2019.1634140

Immergluck D, Smith G (2006) The impact of single-family mortgage foreclosures on neighborhood crime. Hous Stud 21(6):851–866. https://doi.org/10.1080/02673030600917743

ISPRA (Istituto Superiore per la Protezione e la Ricerca Ambientale) (2015) Consumo di suolo in Italia. Rapporti 218/2015. Available at: http://www.isprambiente.gov.it. Accessed 18 Nov 2018

ISTAT (2014) Edifici e abitazioni [Buildings and homes]. https://www.istat.it. Accessed 21 Feb 2019

Jeon Y, Kim S (2020) Housing abandonment in shrinking cities of East Asia: case study in Incheon, South Korea. Urban Studies 57(8):1749–1767. https://doi.org/10.1177/0042098019852024

Kasper W, Streit ME (1998) Institutional economics: social order and public policy. Edward Elgar, Cheltenham

Kraut DT (1999) Hanging out the no vacancy sign: eliminating the blight of vacant buildings from urban areas. New York Univ Law Rev 74(4):1139–1177

Lai LWC (2005) Neo-institutional economics and planning theory. Plan Theory 4(1):7–19. https://doi.org/10.1177/1473095205051437

Latour B (2005) Reassembling the social: an introduction to actor-network-theory. Oxford University Press, Oxford

Legambiente, CRESME (2013) ON-RE Osservatorio Nazionale Regolamenti Edilizi per il risparmio energetico. Available at: https://www.legambiente.it. Accessed 12 Sept 2019

Lieb RC, Merel RA, Perlin AS, Sadoff MB (1974) Abandonment of residential property in an urban context. DePaul Law Rev 23(3):1186–1224

Lieto L (2017) How material objects become urban things? City 21(5):568–579. https://doi.org/10.1080/13604813.2017.1374782

Madanipour A (1999) Why are the design and development of public spaces significant for cities? Environ Plann B Plann Des 26(6):879–891

Mallach A (2012) Depopulation, market collapse and property abandonment: surplus land and buildings in legacy cities. In: Mallach A (ed) Rebuilding America's legacy cities: strategies for cities losing population. American Assembly, New York, pp 85–110

MEF (Ministero dell'Economia e delle Finanze) (2017) Rapporto sul debito pubblico [Public debt report]. Available at: http://www.dt.tesoro.it. Accessed 8 April 2019

MEF (Ministero dell'Economia e delle Finanze) (2018) Rapporto sui beni immobili delle Amministrazioni Pubbliche. Dati 2016 [Report on real estate of public administrations, 2016 data]. Available at: http://www.dt.tesoro.it. Accessed 10 April 2019

MEF (Ministero dell'Economia e delle Finanze) (2019) Gli immobili in Italia. Ricchezza, reddito e fiscalità immobiliare. Available at: https://www.agenziaentrate.gov.it. Accessed 21 Jan 2020

Micclli E, Pellegrini P (2017) Wasting heritage. The slow abandonment of the Italian historic centers. J Cult Herit. https://doi.org/10.1016/j.culher.2017.11.011

Micelli E, Pellegrini P (2019) Paradoxes of the Italian historic centres between underutilisation and planning policies for sustainability. Sustainability 11(9):2614. https://doi.org/10.3390/su11092614

Morckel V (2014) Predicting abandoned housing: does the operational definition of abandonment matter? Community Dev 45(2):121–133. https://doi.org/10.1080/15575330.2014.892019

Morgan DJ (1980) Residential housing abandonment in the United States: the effects on those who remain. Environ Plan A 12(12):1343–1356. https://doi.org/10.1068/a121343

Moroni S (2010) An evolutionary theory of institutions and a dynamic approach to reform. Plan Theory 9(4):275–297. https://doi.org/10.1177/1473095210368778

Moroni S (2015) Complexity and the inherent limits of explanation and prediction. Plan Theory 14(3):248–267. https://doi.org/10.1177/1473095214521104

Moroni S (2018) Property as a human right and property as a special title. Rediscussing private ownership of land. Land Use Policy 70(Jan):273–280. https://doi.org/10.1016/j.landusepol.2017.10.037

Moroni S (2019) Constitutional and post-constitutional problems: reconsidering the issues of public interest, agonistic pluralism and private property in planning. Plan Theory 18(1):5–23. https://doi.org/10.1177/1473095218760092

Moroni S (2020) The just city. Three background issues: institutional justice and spatial justice, social justice and distributive justice, concept of justice and conceptions of justice. Plann Theory 19(3):251–267. https://doi.org/10.1177/1473095219877670

Moroni S, Cozzolino S (2020) Conditions of actions in complex social-spatial systems. In: de Roo G, Yamu C, Zuidema C (eds) Handbook on planning and complexity. Edward Elgar, Cheltenham, pp 186–202

Moroni S, Minola L (2019) Unnatural sprawl: reconsidering public responsibility for suburban development in Italy, and the desirability and possibility of changing the rules of the game. Land Use Policy 86(July):104–112. https://doi.org/10.1016/j.landusepol.2019.04.032

Moroni S, De Franco A, Bellè BM (2020) Vacant buildings. Distinguishing heterogeneous cases: public items versus private items; empty properties versus abandoned properties. In: Lami IM (ed) Abandoned buildings in contemporary cities: smart conditions for actions. Springer, Cham, pp 9–18

Moroni S, De Franco A, Bellè BM (2020) Unused private and public buildings: re-discussing merely empty and truly abandoned situations, with particular reference to the case of Italy and the city of Milan. J Urban Aff 42(8):1299–1320. https://doi.org/10.1080/07352166.2020.1792310

Németh J, Langhorst J (2014) Rethinking urban transformation: temporary uses for vacant land. Cities 40:143–150. https://doi.org/10.1016/j.cities.2013.04.007

North D (1990) Institutions, institutional change and economic performance. Cambridge University Press, Cambridge

Pennington M (2003) Land use planning: public or private choice? Econ Aff 23(2):10–15

Popper K (1974) Unended quest: an intellectual autobiography. Open Court, Chicago

PWC (PricewaterhouseCoopers) (2020) The Italian NPL market. Ready to face the crisis. Available at: https://www.pwc.com. Accessed 24 Nov 2020

Rauws W, Cozzolino S, Moroni S (2020) Framework rules for self-organizing cities: introduction. Environ Plann b: Urban Anal City Sci 47(2):195–202

Salet W (2018) Public norms and aspirations: the turn to institutions in action. Routledge, London

Salet W (ed) (2018) The Routledge handbook of institutions and planning in action. Routledge, London

Sampson N, Nassauer J, Schulz A, Hurd K, Dorman C, Ligon K (2017) Landscape care of urban vacant properties and implications for health and safety: lessons from photovoice. Health Place 46(May):219–228. https://doi.org/10.1016/j.healthplace.2017.05.017

Samsa MJ (2008) Reclaiming abandoned properties: using public nuisance suites and land banks to pursue economic redevelopment. Cleveland State Law Rev 56(1):189–232

Savini F, Majoor S, Salet W (2015) Dilemmas of planning: intervention, regulation, and investment. Plan Theory 14(3):296–315

Schilling J (2009) Code enforcement and community stabilization: the forgotten first responders to vacant and foreclosed homes. Albany Govern Law Rev 2:101–162

Searle J (1995) The construction of social reality. The Free Press, New York. La costruzione della realtà sociale. Einaudi, Torino, 2006

Searle J (2010) Making the social world; Italian translation Creare il mondo sociale, Giuffrè, Milano

Silverman RM, Yin L, Patterson KL (2013) Dawn of the dead city: an exploratory analysis of vacant addresses in Buffalo, NY 2008–2010. J Urban Aff 35(2):131–152. https://doi.org/10.1111/j.1467-9906.2012.00627.x

Smith ME (2020) The comparative analysis of early cities and urban deposits. J Urban Archaeol 2:197–205

Smith ME (2020) Definitions and comparisons in urban archaeology. J Urban Archaeol 1:15–30

Sorensen A (2015) Taking path dependence seriously: an historical institutionalist research agenda in planning history. Plan Perspect 30(1):17–38

Sternlieb G, Burchell RW, Hughes JW, James FJ (1974) Housing abandonment in the urban core. J Am Plann Assoc 40(5):321–332. https://doi.org/10.1080/01944367408977488

Stone CN (1985) Efficiency versus social learning: a reconsideration of the implementation process. Rev Policy Res 4(3):484–496. https://doi.org/10.1111/j.1541-1338.1985.tb00248.x

vom Hofe R, Parent O, Grabill M (2019) What to do with vacant and abandoned residential structures? The effects of teardowns and rehabilitations on nearby properties. J Reg Sci 59(2):228–249. https://doi.org/10.1111/jors.12413

Wang K, Immergluck D (2019) Housing vacancy and urban growth: explaining changes in long-term vacancy after the US foreclosure crisis. J Housing Built Environ 34(2):511–532. https://doi.org/10.1007/s10901-018-9636-z

White MJ (1986) Property taxes and urban housing abandonment. J Urban Econ 20(3):312–330. https://doi.org/10.1016/0094-1190(86)90022-7

Wilson D, Margulis H, Ketchum J (1994) Spatial aspects of housing abandonment in the 1990s: the Cleveland experience. Hous Stud 9:493–510

World Bank (2019) Doing business 2019. World Bank, Washington (DC). Available at: www.doi ngbusiness.org. Accessed 19 June 2020

WWF (2013) Riutilizziamo l'Italia. Dal censimento del dismesso scaturisce un patrimonio di idee per il futuro del Belpaese, Edizioni WWF. Available at: https://wwfit.awsassets.panda.org. Accessed 3 April 2018

Zhang S, de Roo G, Rauws W (2020) Understanding self-organization and formal institutions in peri-urban transformations: a case study from Beijing. Environ Plann b: Urban Anal City Sci 47(2):287–303

Chapter 2
Theoretical Framework

Abstract This section is divided into two main parts. The first exposes the main theoretical assumptions, in particular the distinction between "brute facts" and "social facts" (Sect. 2.1). Considering "abandonment as a social fact" is a thesis in *social ontology* rather than, in itself, a *theory in sociology*; it precedes any eventual sociological discussion. The second discusses the differences between functioning, deteriorated, empty and abandoned buildings (Sect. 2.2). The idea is to suggest a general theory of abandonment as a *social fact*, according to which abandonment is a potential state of *any* urban asset.

Keywords Brute facts · Social facts · Responsibility · Ownership · Pragmatic duty · Social ontology

2.1 Abandonment as a Social Fact

Here, we define as "abandoned" a building whose *owner* does not fulfil his/her *responsibilities* ensuing from ownership.[1] According to this view, abandonment is considered a *social fact*. To clarify what is meant here by "social fact", it is useful to distinguish between "brute facts" and "social facts".

1. A "brute fact" is a fact that exists independently from human acceptance, agreements or institutions (Searle 2010).[2] For example, building A is X away from building B.

---

[1] A similar view can be found for instance in Kraut (1999, p. 1140), Hillier et al. (2003, p. 93), Mallach (2006, p. 1), Han (2019, pp. 3–4).

[2] The expression "brute fact" has been employed by Gertrude Elisabeth Margaret Anscombe, Charles Sanders Peirce, Henri Poincarè, and subsequently, made popular by Searle (1995). See on this Lorini (2000, pp. 54 ff.). Note that brute facts are the "inhabitants" of Popper's *World 1* (Popper and Eccles 1977).

2.   A "social fact" is instead and conversely, a fact that exists only through (because of/by reason of) human acceptance, agreements or institutions. For example, W is the mayor of city Z.[3]

Durkheim (1895/1982) was one of the first to use the concept of "social fact" in this line of thinking. Social facts in this sense are something more than simple psychic facts existing in individual consciences (Fig. 2.1).[4]

Abandonment is a social fact because at least two conventional/institutional concepts are required to define it: that of "owner" and that of "responsibility".[5]

Fig. 2.1   Social facts beyond single minds. Author's elaboration from Searle (1995)

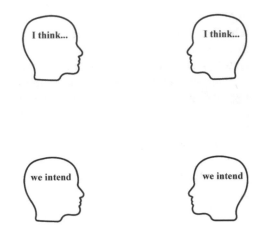

---

[3] In this particular expanded meaning of the idea of "social fact", what Searle (1995) calls "institutional facts" can be considered as a subclass of social facts (Searle 2006, p. 17) that exist as a consequence of a *declaration* (Searle 2010): for example, "the session of the city council is open".

[4] As Durkheim (1895/1982, p. 53) writes, with reference to social facts: "Here, then, is a category of facts which present very special characteristics: they consist of manners of acting, thinking and feeling external to the individual. … Consequently … they cannot be confused with organic phenomena, nor with psychical phenomena, which have no existence save in and through the individual consciousness. Thus they constitute a new species and to them must exclusively assigned the term *social*" (emphasis in the original). Against *methodological psychologism*, see also Popper (1945, pp. 301–310).

[5] In Searle's term (2010) a "deontology" is implied here. Compare with the case of the abandonment of vulnerable subjects (minors, but also pets). This issue is often disciplined by penal codes (in Italy, for instance by the art. 591 and art. 727, *Codice Penale*). The *ratio legis* is that abandonment is an offence or mistreatment of living beings; by reason of "evolution of costumes and social understandings [*istanze sociali*]" (see the sentence no. 6317 of 2004 by the Italian *Consiglio di Stato*). In Italy, as Grosso (2019) points out, jurisprudence interpret the abandonment concept in terms of "virtual" or "potential" threat, without attributing concrete boundaries to the concept. This situation makes the imputation of a crime particularly hard to define. While it is clear that the offense exists, in general, with omissions of custody obligations, it is not easy to define when the "guardian" has the precise and definitive intention to abandon. Note how this is also relevant for naval codes (in Italy, art. 1097 and 1098, *Codice Navale*).

The type of responsibility at stake, in this case, could be defined as a "pragmatic duty"; that is, a duty that exists for someone not in an *absolute* sense, but *in relation* to his/her status—in our case, the status of the owner of an asset (Conte 2006; Di Lucia 2008). It is important to emphasize here that the status of the owner, therefore, implies both *rights* and *duties* (Shoked 2014). This is even truer in urban contexts where properties are concentrated.[6] With particular reference to these situations, we assume that social agents cannot renounce their pragmatic duties (Fig. 2.2).[7]

Note that a *building* (as a *brute fact*) can be directly perceived with the senses and, for example, photographed; *abandonment* (as a *social fact*) cannot instead be directly perceived with the senses and, for example, photographed because it is a social/conventional phenomenon. (The point is also valid when speaking of "abandoned minors").[8] Note also how, from the perspective adopted here, the degraded/dilapidated condition of a building is a "sign" of its possible state of abandonment, not *abandonment* as such.

It has been discussed in the literature whether "brute facts" and "social facts"/"institutional facts" are always present together: for example, Searle (1995), argues that institutional facts exist only "on the top of brute physical facts", while according to Lorini (2000, pp. 64 ff.) not all social or institutional entities always need a physical substrate. Regardless of this more general debate, in our case—that is, the issue of abandoned buildings—the social and brute facts are always *co-present*: they cannot be separated.

The distinction between "brute facts" and "social facts" could also be made by considering the possible non-linguistic and linguistic reactions. Degradation, as a brute fact, can generate a sense of fear, while abandonment, as a social fact, can generate linguistic reactions (e.g. "What a waste!", "It is so unjust!"). This happens because, in the first case, one reacts to a "visible ontology", while in the second case, the reaction descends from an "invisible ontology" (Searle 1995, 2010).

Note that recognising "abandonment as a social fact" is, first of all, a thesis in *social ontology* rather than, in itself, a *theory in sociology*: it precedes any eventual

---

[6] It is interesting to observe that even a classical liberal like von Hayek (1960, p. 341) recognizes this: "the costs involved in large numbers living in great density not only are very high but are also to a large extent communal, i.e. they do not necessarily or automatically fall on those who cause them but may have to be borne by all. In many respects, the close contiguity of city life invalidates the assumptions underlying any simple division of property rights. In such conditions it is true only to a limited extent that whatever an owner does with his property will affect only him and nobody else". And he goes on by stressing that: "The general formulas of private property or freedom of contract do not therefore provide an immediate answer to the complex problems which city life raises".

[7] A different issue which is not discussed in depth here is whether social agents can renounce their status (e.g. being an owner).

[8] For instance, if you see a child sitting or walking alone, how can one be sure/know whether the kid is abandoned or not? As Conte (2011, pp. 27–34) stresses, the same behavior or action may be in light of different background rules. As a consequence, by the mere observation of a behaviour or action we cannot directly derive/infer the rules behind them.

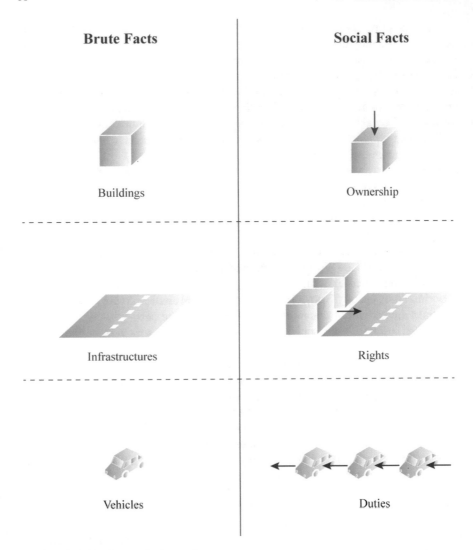

Fig. 2.2 Building blocks of urban reality

sociological discussion.[9] In other words, the main point here is to explore and clarify what the foundations of any sociological discourse on abandonment are.

---

[9] This approach could be understood as an attempt to enter the study of the "philosophy of society". According to some authors, this branch originated from Durkheim's conceptions of society as "collective representation" (Gafijczuk 2005). Searle (2006, 2010) makes explicit reference to it alongside other disciplines such as the philosophy of mind and language. Other authors underlined how Searle's realist approach can be put in contrast with pragmatist and other sociological approaches (Kivinen and Piiroinen 2007). The scope of this work is not to further stretch such arguments, but to try to reflect on social ontology to reconsider the relationships and interactions

2.2 Functioning, Deteriorated, Empty and Abandoned Buildings

In scientific literature, there are various definitions of "abandoned buildings", just as there are many ways to parallel abandonment problems to other undesirable processes in the built environment (urban shrinkage, decay, decline and so on). The lack of shared definitions unveils the conceptual challenges in understanding abandonment both as an end-state of structures and as evidence of socioeconomic changes in urbanized areas.

In this study, an "abandoned building" is defined as:

> A built structure that is not fully serviceable and habitable in its present state because of its advanced state of functional and physical degradation, due to the fact that the building's owner fails to fulfil his/her responsibility (i.e. pragmatic duty) ensuing from ownership.

Note that, in this light, the distinction between a truly "abandoned" building and a merely "empty" building is clearer. Both are in fact "not occupied/used", but in the first case, the owners do not maintain the structures, while in the second this does happen.[10]

The above definition stresses some aspects of the "abandoned state" of buildings that are often overlooked although crucial: (i) abandoned buildings are antithetical to fully *functioning buildings*; (ii) abandoned buildings are a sub-group of *unused buildings*; (iii) abandoned buildings are different from occupied but *deteriorated buildings*.

There can be causal relationships between occupation and maintenance among building states (i.e. functioning, empty, deteriorated, abandoned buildings), but literature tends to overlook how these relate dynamically to each other (Buitelaar et al. 2021). For instance, when emptiness and poor maintenance coincide, this would lead to abandoned buildings (Table 2.1).[11]

Table 2.1 Four archetypical building states based on Buitelaar et al. (2021)

		Management	
		Yes *(well-maintained)*	No *(unmaintained)*
Occupation	Yes *(used)*	Functioning	Deteriorated
	No *(unused)*	Empty	Abandoned

---

between human agents and their contexts of action. For similar attempts, see Anderson (2012) and Smith (2020).

[10] Obviously, it could be argued that the (pragmatic) duties of an owner must be broader, but it seems difficult to include among the duties of a good owner to keep, for example, the building he/she owns always occupied or managed by someone.

[11] Consequently, if no further action takes place, abandoned buildings will become *derelict*.

Observe how it seems difficult to discuss abandonment only in temporal terms, even in relation to technology cycles.[12] This is due to the fact that *time* in itself is a *brute fact*.

It is, therefore, possible to conceive a general theory of abandonment (as a *social fact*) according to which abandonment is a potential state of *any* property that is directly dependent on *human actions/decisions* (e.g. use and maintenance).

It is interesting to remember here that in Roman law there was a sharp distinction regarding dereliction: that between material elements (i.e. a "*corpus*") and human elements (i.e. an "*animus*"). The general assumption was that for every abandoned thing (i.e. *res derelictae*) there was a relinquishing subject (to whom an *animus derelinquendi* could be attributed); he/she is physically separated from the object in his/her possession, acting as if he/she is no longer interested in using it (Romano 2002, p. 4; Gambaro and Morello 2011).

The relationship between material and social elements has been reinterpreted during the centuries, giving more prominence to "brute" aspects (relinquishment *of objects*) while overlooking "social" ones (i.e. abandonment as an *intention*; Romano 2002, pp. 102–106).

Focusing attention on building assets, issues such as "dereliction" do not automatically attest to a voluntary loss of possession and therefore, it does not per se substantiate the will to lose property rights in general terms.[13] Going back to the concept of pragmatic duty and the presence of a special/conventional ontology of property assets, it seems possible to say that abandoned buildings lose their agentive functions (they are neither used nor usable by social agents).

References

Anderson J (2012) Relational places: the surfed wave as assemblage and convergence. Environ Plann d: Soc Space 30(4):570–587

---

[12] Temporal variables such as months or years of (non)usage may inadvertently push one to consider abandoned buildings and other sub-standard properties altogether. See on this Merrett and Smith (1986), Dewar and Weber (2012), Dewar et al. (2015), Raleigh and Galster (2015), Wang and Immergluck (2019), Jeon and Kim (2020). On the issue of time in ontological studies, see Jacquette (2002).

[13] Also in Italy, as in many other countries of continental Europe, there is a discussion whether "wasted" buildings (or goods in general) are automatically acquired by governments (see on this Resta 2018, pp. 255–258) once the owner has ceased to fulfil the relative responsibilities. The aim of this work is not to add or develop any discussion around these matters, also to avoid conceptual and operative confusion. Here we assume that even if the state of abandonment testifies possible problems and shifts in the property domain, they remain "private properties" unless expropriation or acquisition takes place.

Buitelaar E, Moroni S, De Franco A (2021) Building obsolescence in the evolving city. Reframing property vacancy and abandonment in the light of urban dynamics and complexity. Cities 108 (online ahead of print)

Conte AG (2006) Dovere pragmatico secondo Epicuro. In: Limone G (ed) L'arcipelago dei diritti fondamentali alla sfida della critica. Angeli, Milano, pp 56–61

Conte AG (2011) Sociologia filosofica del diritto. Giappichelli, Torino

Dewar M, Weber MD (2012) City abandonment. In: Dewar M, Weber MD (eds) The Oxford handbook of urban planning. Oxford University Press, Oxford, pp 563–586

Dewar M, Seymour E, Druță O (2015) Disinvesting in the city: the role of tax foreclosure in Detroit. Urban Aff Rev 51(5):587–615

Di Lucia P (2008) Doveri pragmatici in Cicerone. In: Lanchester F, Serra T (eds) Et si omnes... Scritti in onore di Francesco Mercadante. Giuffrè, Milano, pp 283–296

Durkheim É (1895) Les règles de la méthode sociologique. Alcan, Paris; The rules of sociological method. The Free Press, New York, 1982

Gafijczuk D (2005) The way of the social: from Durkheim's society to a postmodern sociality. Hist Hum Sci 18(3):17–33

Gambaro A, Morello U (2011) Trattato dei diritti reali (vol 1). Giuffrè, Milano

Grosso S (2019) (Meta)paternalismo giudiziale e abbandono di minori. Diritto Penale Contemporaneo 7–8:65–73

Han HS (2019) Exploring threshold effects in the impact of housing abandonment on nearby property values. Urban Aff Rev 55(3):772–799. https://doi.org/10.1177/1078087417720303

Hayek FA (1960) The constitution of liberty. Routledge, London

Hillier AE, Culhane DP, Smith TE, Tomlin CD (2003) Predicting housing abandonment with the Philadelphia neighborhood information system. J Urban Aff 25(1):91–106. https://doi.org/10.1111/1467-9906.00007

Jacquette D (2002) Ontology. Acumen, Chesham

Jeon Y, Kim S (2020) Housing abandonment in shrinking cities of East Asia: case study in Incheon, South Korea. Urban Studies 57(8):1749–1767. https://doi.org/10.1177/0042098019852024

Kivinen O, Piiroinen T (2007) Sociologizing metaphysics and mind: a pragmatist point of view on the methodology of the social sciences. Hum Stud 30(2):97–114

Kraut DT (1999) Hanging out the no vacancy sign: eliminating the blight of vacant buildings from urban areas. New York Univ Law Rev 74(4):1139–1177

Lorini G (2000) Dimensioni giuridiche dell'istituzionale. Giuffrè, Milano

Mallach A (2006) Bringing buildings back: from abandoned properties to community assets. Rutgers University Press, New Brunswick (NJ)

Merrett S, Smith R (1986) Stock and flow in the analysis of vacant residential property. Town Plann Rev 57(1):51–67

Popper K (1945) The open society and its enemies. Routledge, London

Popper K, Eccles J (1977) The self and its brain. Springer, Berlin

Raleigh E, Galster G (2015) Neighborhood disinvestment, abandonment, and crime dynamics. J Urban Aff 37(4):367–396

Resta G (2018) I rifiuti come beni in senso giuridico. Rivista Critica Del Diritto Privato 36(2):249–268

Romano S (2002) Studi sulla derelizione nel diritto romano. Rivista Di Diritto Romano 2:99–164

Searle J (1995) The construction of social reality. The Free Press, New York. La costruzione della realtà sociale. Einaudi, Torino, 2006

Searle J (2006) Social ontology: some basic principles. Anthropol Theory 6(1):12–29

Searle J (2010) Making the social world; Italian translation Creare il mondo sociale. Giuffrè, Milano

Shoked N (2014) The duty to maintain. Duke Law J 64(3):437–513

Smith ME (2020) Definitions and comparisons in urban archaeology. J Urban Archaeol 1:15–30

Wang K, Immergluck D (2019) Housing vacancy and urban growth: explaining changes in long-term vacancy after the US foreclosure crisis. J Housing Built Environ 34(2):511–532. https://doi.org/10.1007/s10901-018-9636-z

Chapter 3
Analytical Schema

Abstract In this sub-section and once recognized abandonment as a social fact, an analytical schema based on *parameters* (Sect. 3.1), *factors* (Sect. 3.2), *reasons* (Sect. 3.3), *policies* (Sect. 3.4) is proposed. The aim is to have a richer "toolbox" to critically revisit the issue as a complex problem.

Keywords Parameters · Indicators · Factors · Reasons · Policies

3.1 Parameters and Indicators

There are at least five operative parameters (and relative indicators) to detect "abandoned buildings": *physical, facility, fiscal, economic, statutory* parameters. Some of them (i.e. physical and facility indicators) are mere "signs" of possible states of abandonment—they reflect brute facts—while others try to identify abandonment more substantially as a social fact (i.e. fiscal, economic, statutory).[1] Obviously, these indicators can be used in combination. Let us consider them in detail.

1. *Physical parameters*, for example, the peeling of façade paint, deterioration and collapse of the structures, signs of vandalization, infestations, etc. deriving

---

[1] In this regard, Searle (1995, p. 85) points out: "possession is typically manifested quite differently for real property than for personal property. I can wear my shirt, drive my car, even carry my computer, but when it comes to my house and land, maintenance of my possession requires status indicators". He goes on underlying that: "The French distinction between '*meuble*' and '*immeuble*' reveals precisely this distinction" (Searle 1995, p. 84; italics in the original). This similarly applies in the Italian language (i.e. "*mobili*", "*immobili*").

© The Author(s), under exclusive license to Springer Nature Switzerland AG 2022   21
A. De Franco, *Abandonment as a Social Fact*, SpringerBriefs in Geography,
https://doi.org/10.1007/978-3-030-90367-1_3

from the neglect of the assets.[2] These elements can be assessed via on-the-spot observation mainly of the exterior building structure.[3]

2. *Facility parameters*, that is, termination of basic utilities and usage level, for example, electricity or water.[4] This information can be available for public subjects but not necessarily for every subject because of privacy and/or costs to assess this kind of information.
3. *Fiscal parameters*, for example, uncollected municipal taxes or rents and foreclosures etc. They are "classical" indicators of abandonment (Sternlieb et al. 1974; White 1986; Morckel 2014).[5] This information is available in local property records and registries.
4. *Economic parameters*, for example, market values. They can be useful to understand price development for the structures which usually depreciate in their life cycle (Baum 1991, p. 59), especially if no intervention takes place.[6]

---

[2] Typically, these types of indicators are linked, at least in origin, to a pure technical language used by administrative offices and technical inspectors (e.g. planning and building inspectors, tributary offices, fire brigades, police). In the latest Italian census, Istat (2014) estimates that 354,848 units of the national building stock are "buildings unsuitable for productive use or income, due to the accentuated level of degradation". According to estimates from other institutes and trade associations (e.g. Confedelizia, Unimpresa), in 2017, there were half a million of these properties included in the "derelict" category (in 2017, 452,410 units, equal to 0.72% of the global stock). The data have been updated to 2015 based on analysis of the Land Registry and the Court of Auditors (i.e. Corte dei Conti). See https://www.unimpresa.it/edilizia-unimpresa-mezzo-miliione-di-immobili-in-dissesto/15184 (Accessed June 2019).

[3] Thus, it is difficult and discouraged to break into the property since it is "imminently dangerous" (Hillier et al. 2003, pp. 93–94). As also noted by Kraut (1999, p. 1148) even police or other civil servants may be reluctant to enter these properties.

[4] Note how, even in this case, the indicators do not simply record brute facts but social practices, also in terms of use and performance of local services (e.g. detecting uncollected waste, unanswered mails). Thus, some municipalities detect vacant/abandoned properties by the rate of response to municipal warrants. Whenever the property does not reply to the solicitations, governments may classify certain properties as a sort of "undeliverable addresses" (Schilling 2009, p. 108; Silverman et al. 2013; Molloy 2016; Newman et al. 2018).

[5] Fiscal indicators are considered as an accurate evidence of the cease of concrete interests in the use or possession of the property. They can be useful to assess *decommissioned* buildings and may signal different processes in place. Decommissioning a property may be because of concrete suffering of economic agents (e.g. physical persons, as local owners or firms) or part of a more general liquidation strategy of financial operators (e.g. non physical persons, as capital funds or credit institutions; PWC 2020). For specific reference on the situation of private owners and construction sector in Italy, see ANCE (2012, 2019).

[6] In this section we talk about indicators that detect the abandonment of buildings. In the case of economic indicators, of course, it is necessary to take account of the fact that the decrease in the value of the building does not correspond to a decrease in the value of the overall property. This happens because the value of the land can remain or even rise, especially in vital urban contexts, regardless of the deterioration of the structures on the site. Notably, public bodies also have a role in this price inflation in land markets when creating—or not—public amenities (Bowman and Pagano 2000, p. 560; Kelly 2013, pp. 117–118). Only in certain circumstances land values may decrease. For instance, when nuisances are pervasive to the whole urban compounds (e.g. because of fire accidents; see Morgan 1980; Kraut 1999, p. 1150) and, allegedly, in shrinking contexts.

5. *Statutory parameters*, that is, when abandoned buildings are formally registered as such, for instance, through formal "declarations".[7] This last criterion cannot be taken as absolute, as definitions may vary depending on language and legal doctrines.[8]

The operative elements here considered (Table 3.1) identify abandonment problems in a generic but extensive manner: the parameters/indicators can be useful to control essential actions taken (or not taken) *through* time by the owners.[9] Obviously, some of them need to be readjusted as empirical processes constantly challenge their validity. For instance, some owners may normally pay their taxes even if the property lies derelict to avoid costly fines or litigation with local governments (Kraut 1999, p. 1152; Abe et al. 2014, p. 351). This situation reveals the need to adjust certain information by combining the indicators.

Table 3.1 Abandonment parameters and indicators

Parameter	Indicator	Method	Accessibility
Physical	Signs of deterioration, vandalization, collapse of the structures, accumulation of debris, growing vegetation, etc.	On-the-spot observation	To everyone
Facility	Termination of basic *utilities* and *usage* level, for example, electricity, water, etc.	Providers' records	Not to everyone for reasons of privacy
Fiscal	Uncollected municipal taxes, foreclosures etc.	Property records	Public officials
Economic	Market values	Price records	Mainly to market operators but potentially to everyone
Statutory	Formal declarations	Public records	To everyone

---

[7] In rare cases, and in Italy as well, the abandoned state is declared through a notarial act that virtually transfers derelict properties to other subjects (as *"res nullius"* or *"derelictae"*—"abandoned things"—Civil code, art. 586, 711, 812). These sometimes become "State assets" under a traditional assumption of the State as "principal owner" (i.e. *Stato come primo soggetto proprietario*, Civil code of 1865, art. 425). Here, it is important to note the differences in abandonment issues in private and public cases (Moroni et al. 2020a, b).

[8] As Schilling (2009, p. 161) writes: "statutory language [is] not sufficiently sensitive to the important difference between weak and strong market cities". See on this Keenan et al. (1999), Morckel (2014), Haase et al. (2014, p. 1521). Note how, based purely on linguistic expressions, theories run the risk of being "circular". This is one of the reasons why in this work we suggest interpolating multiple indicators to identify abandonment problems.

[9] For Popper (1974, pp. 145–147), humans do not have full experience of spatial and temporal coordinates. The use of temporal variables in abandonment studies (e.g. years, months of non-use of assets) are merely conventional (Merrett and Smith 1986, pp. 63–64). Time by itself does not help us to understand/explain the phenomenon; it is an independent parameter. See also Jacquette (2002, p. 195).

Table 3.2 Abandonment factors

Factors	Description
Psychological factors (individual attitudes)	*Negative game* i.e. unwilling, disinterested owner
	Positive game i.e. speculation, strategic owner
Economic factors (global market dynamics)	*Crisis in the building sector* i.e. construction and real estate markets, financial crisis
	Crisis in property markets i.e. private wealth
Procedural factors (institutional)	*Regulatory and bureaucratic barriers* i.e. processual constraints
Functional factors (typological and locational)	*Building features* i.e. size, shape, position

Note that, considering abandonment as a social fact, the role of the social agent becomes prominent in explaining that abandonment can be conceived both as (i) a *behaviour* of those who abandon something; (ii) a *decision*, a sort of voluntary loss.[10]

3.2 Factors of Abandonment

The aim of this sub-section is to identify *why* certain buildings are abandoned. The focus will be on four main factors: *psychological, economic, procedural* and *functional*. Abandonment factors are conceived here as those elements that drive, trigger or condition the behaviour of social agents. In other words, they are not interpreted as causes in a deterministic, linear way, but as background conditions that may influence certain outcomes. Obviously, the four factors are not alternative as they can be cumulative and simultaneously occur in urban contexts. Let us consider them in detail (Table 3.2).

1. *Psychological factors.* Single owners' mindsets may influence abandonment states.[11] There are two main recurrent personal attitudes. A first possibility is when the owner plays a "negative game"; that is when owners are unwilling to use or upgrade their assets.[12] A second possibility is when the owner

---

[10] See Sternlieb et al. (1974) and Basset et al. (2006). Also Ross and Portugali (2018).

[11] As Edson (1972, p. 382) emphatically stresses: "the absolute number of abandoned structures is much less important than the process of abandonment, and the reasons and the state of mind that have produced it". In the same line, see also Sternlieb et al. (1974, pp. 326–329), Lieb et al. (1974, pp. 1193–1195), Morgan (1980). Other factors mentioned in the literature (e.g. reputation/stigmatization of certain areas; see Galster 2001) may be subsumed in these general analytical categories.

[12] See Lieb et al. (1974, p. 1193), Sternlieb et al. (1974), Keenan et al. (1999, p. 705), Basset et al. (2006, pp. 58–59), Henderson (2015,) Holtslag-Broekhof (2018).

plays a "positive game"; that is when owners deliberately postpone their decisions waiting for profitable revenues in the future (typically for reasons of speculation).[13]

2. *Economic factors*. Particularly in the light of the latest financial crisis, the accumulation of stalled projects, forced sales, etc. consolidated the classical relation between abandonment and economic factors. Abandoned buildings may be considered detached from local markets (Keenan et al. 1999, p. 705). Decreasing property values both signal and impinge on the stagnation of the building sector and property markets.[14]

3. *Procedural factors*. Procedural factors are those formal "constraints" on the possibility of bringing the property back into market circuits (for the exchange and/or transformation of the property: Merrett and Smith 1986; Accordino and Johnson 2000; Schilling 2009; Holtslag-Broekhof 2018; Garda 2018). These factors depend on too strict or costly prescriptions, typically enforced by planning and governance agencies that control urban densification, and on connected bureaucratic procedures (Farris 2001; Talen 2012; Hong et al. 2016; Olivadese et al. 2017).

4. *Functional factors*. These regard typologies and locations; the issue here is not geometry or spatial coordinates (brute facts), but, rather, design and place (social facts). The first concern here is to understand whether certain types of properties could have different uses in the future. Considering issues of design adaptation, certain buildings typologies are more difficult to transform.[15] For instance, tower buildings typically host multiple and various units in a rigid way and in the same vertical structure (Wegmann 2020), while it is easier to intervene on buildings that are not excessively tall and with open and flexible interiors and adaptable structures (e.g. industrial hangars: Remøy 2010; Remøy and Street 2018; Lami 2020). The second concern, considering building locations, is how contextual features influence abandonment processes. This is not exclusively related to

---

[13] See Lieb et al. (1974, p. 1197), White (1986), Kraut (1999, p. 1152), Bowman and Pagano (2000, pp. 561), Glock and Häußermann (2004, p. 924), Németh and Langhorst (2014, p. 144), Galster (2019, pp. 357–358).

[14] In this regard, some authors go more in depth discussing on (i) the construction sector (Power and Mumford 1999; Accordino and Johnson 2000; Glaeser and Gyourko 2005; Immergluck and Smith 2006; Harding et al. 2009; Bogataj et al. 2016; Wang and Immergluck 2019), (ii) private property markets (Glock and Häußermann 2004; Schilling 2009; Campbell et al. 2011; Mallach 2012; Whitaker and Fitzpatrick 2013; Raleigh and Galster 2015; Cozens and Tarca 2016), and (iii) regulatory barriers and constraints (Edson 1972, p. 385; Lieb et al. 1974, p. 1205; Merrett and Smith 1986, p. 57; Kraut 1999; Accordino and Johnson 2000, pp. 310; Bowman and Pagano 2000; Cohen 2001, p. 437; Farris 2001; Hillier et al. 2003, pp. 94–95; Basset et al. 2006, pp. 58–59; Mallach 2006, 2012, p. 104; Schilling 2009, p. 115; Fuentes et al. 2010; Talen 2012; Silverman et al. 2013; Henderson 2015; Németh and Langhorst 2014; Hong et al. 2016; Olivadese et al. 2017; Garda 2018; Holtslag-Broekhof 2018; Humphris and Rauws 2020).

[15] See Sternlieb et al. (1974, p. 322), Taylor et al. (1992), Brand (1995), Remøy (2010, pp. 36), Remøy and Street (2018), Thomsen and Van der Flier (2011), Mallach (2012, p. 94), Grover and Grover (2015, pp. 305–30), Hughes and Jackson (2015, p. 240), Wallace and Schalliol (2015), Han (2017, p. 840), Olivadese et al. (2017), Lami (2020), Wegmann (2020), Humphris and Rauws (2020).

physical/quantitative elements (e.g. position), as prospects for investments may depend on the quality of public areas and amenities (e.g. maintenance of open spaces, accessibility levels, public infrastructures).[16]

3.3   Reasons Advanced to Identify Abandonment as a Negative Phenomenon

The abandonment problem is particularly challenging and multifaceted in urban contexts; technicians, academics, and social advocates bring many issues under public attention.[17] This sub-section considers four main arguments usually advanced to qualify "abandonment" as a negative phenomenon: *nuisance, economic, environmental* and *equity* reasons. To avoid misunderstandings, observe that there is a difference between "nuisances" and "negative externalities". The former negatively affect general and predefined *rights*, while the latter negatively affect specific economic *situations*.[18] Let us consider them in detail (Table 3.3).

---

[16] Abandonment problems tend to be concentrated in non-appetible areas, where market operators cannot compensate for the low-accessibility or low-quality of the surroundings (e.g. urban peripheries or dispersed settlements: Bowman and Pagano 2000; WWF 2013, p. 28; Foster and Newell 2019). In Italy, however, abandonment problems also occur in vital economic regions and in high-quality compact settlements (Micelli and Pellegrini 2019; Adobati and Garda 2018). This, in the view of some of the interviewees, happens because of the viscosity (e.g. redundancy, rigidity, obsolescence) of the rules governing urban transformation (e.g. technical prescriptions, mono-functional zoning: Jacobs 1961, pp. 257–263; Taylor et al. 1992; Brand 1995). See also Buitelaar et al. (2021).

[17] The general criticism is that interventions/policies risk to be "unproductive" (e.g. costs outweigh benefits) while discourses frame the phenomena in mono-directional ways often using fatalist tones. As Keenan et al. (1999, p. 705) emphatically write: "[a]bandoned property not only represents a wasted housing resource but [its] contagion effect …, unless reversed, can lead to whole neighbourhoods becoming devoid of social and economic activity, a twilight zone in which crime and social malaise abound." See also Spelman (1993).

[18] Some authors say that externalities impinge on the reuse potential of the property and represent a social cost for communities (Samsa 2008; Mallach 2012, p. 100). The externality focus is appropriate to describe how disinvestments affect utility functions of nearby owners and occupiers (Basset et al. 2006, pp. 58–59).

Table 3.3 Variety of reasons for identifying abandonment as a negative phenomenon

Variety of reasons	Descriptions
Nuisance i.e. risk to human safety and security	"Attracting disorder and crime to an area, vacant buildings intimidate law-abiding citizens, limiting their activity in the neighbourhood and causing a breakdown in the area's 'natural surveillance' system. … [P]eople who live in distressed neighbourhoods marked by such nuisances as boarded-up buildings or vacant lots report worse health and higher rates of depression, in part because they walk less and isolate themselves inside their homes." (Kraut 1999, pp. 1150–1152)
Economic i.e. negative economic externalities as an effect of disinvestments	"The strategy of minimal maintenance is indeed a dangerous one since, if left uncontrolled, it will lead the owner down the dead end road to hard core disinvestment. … It is at this stage in the abandonment process that the cycle becomes irreversible; the structures have deteriorated to such a degree that further capital reinvestment would be financially unwise. … [T]he absence of prolonged ownership patterns collectively combine to ignite a general negative feeling towards capital investment on the part of landlords, tenants, financial institutions, and government and public agencies." (Lieb et al. 1974, p. 1193) "The durability of the physical stock of structures means that after demand falls and the value of housing and other property declines, disinvestment and deterioration occurs over a long period. During this time, an increasingly rundown building blights its surroundings, undermines the confidence of residents and local business owners in the future of the area, and lowers property values." (Dewar et al. 2015, p. 594)

(continued)

Table 3.3 (continued)

Variety of reasons	Descriptions
Environmental i.e. reuse existing assets to limit new urbanization	"A positive relationship is expected between the amount of vacant land and city policy activism. Given that vacant land represents an unrealized resource, city officials should respond to unused capacity with policies designed to encourage the use and reuse of vacant land." (Bowman and Pagano 2000, p. 563) "The majority of future urban developments would fall under the efforts of reusing existing land and scarce resources, and urban voids have the immanent potential for reuse and redevelopment … Reusing these spaces and resources for human use and to satisfy environmental objectives would incrementally transform the physical and socioeconomic environments of future cities." (Hwang and Lee 2019, p. 10)
Equity i.e. expand the offer of urban structures for those who cannot enter or participate in local markets	"Two principles – efficiency and equity – offer guidance for planners' interventions. Although urban planning as a field has a considerable understanding of why cities lose employment and population and why abandonment occurs, planners could make much more progress … Urban disinvestment and abandonment disproportionately affect poor and minority populations. Therefore, the pursuit of efficiency in land use, housing, and public service delivery in cities with extensive disinvestment and abandonment can affect those residents the most. The challenge for urban planners becomes how to assure equity in this context." (Dewar and Weber 2012, pp. 564–577) "[T]he number of vacant dwellings may be regarded as an indicator of the socioeconomic well-being of a neighbourhood … Recently, the excessive amount of vacant dwellings has been addressed as an issue of social justice in the context of European housing shortage … linking the number of vacant homes to that of the homeless" (Gentili and Hoekstra 2018, p. 4)

1. *Nuisance reasons.*[19] In this case, abandonment is considered undesirable because abandoned buildings represent a risk to human safety and security.[20] As well known, the term "nuisance" is the traditional term for tangible damage to the physical integrity of individuals or other properties: consider the possibility of falling roof shingles, ledges, fire hazards, etc. (Morgan 1980; Jennings 1999; Cohen 2001; Hirokawa and Gonzalez 2010). Certain abandoned properties accumulate in the "edgelands"; being distant and disconnected from urban activities, these remain largely unsurveilled (Humphris and Rauws 2020). Here, but especially in dense areas, it is suggested to have more stringent local codes and stronger enforcement of them (inspections, fines, etc.).

2. *Economic reasons.*[21] In this case, abandonment is considered undesirable because it generates negative economic externalities. Here, the main issue is the value depreciation of the properties surrounding abandoned buildings (Immergluck and Smith 2006; Raleigh and Galster 2015; Han 2019; Galster 2019). Owners of surrounding properties are subject to certain costs for which the owner of the distressed property did not take account (Heller 1998; vom Hofe et al. 2019). Urban planners typically discuss these externalities in order to implement and support redevelopment interventions (Dewar and Weber 2012).

3. *Environmental reasons.*[22] In this case, abandonment is considered undesirable because it seems a waste not to (re)use abandoned assets. As a consequence, property abandonment, decommissioning, or dereliction enters the general debate on urban sustainability and recycling of the built environment (Marini and Santangelo 2013, p. 95). The central claim is usually that the abandonment of urban buildings should be discouraged because instead of using new open land, it is preferable to reuse existing structures and promote densification.[23] In line with this view, various forms of land seizure regulations are inspired by ecological motives and rightsizing strategies at various urban scales (Glock

---

[19] See Morgan (1980), Spelman (1993), Ellickson (1996), Jennings (1999), Kraut (1999), Ross and Mirowsky (1999), Cohen (2001), Farris (2001), Mallach (2006, 2012), Samsa (2008), Schilling (2009), Hirokawa and Gonzalez (2010), Seo and von Rabenau (2011), Silverman et al. (2013), Morckel (2014), Aiyer et al. (2015), Bennett and Dickinson (2015), Raleigh and Galster (2015), Wallace and Schalliol (2015) Sampson et al. (2017).

[20] This does not coincide with merely perceptual problems: e.g. perceived disorder.

[21] See Lieb et al. (1974), Heller (1998), Keenan et al. (1999), Power and Mumford (1999), Accordino and Johnson (2000, p. 301), Farris (2001), Hillier et al. (2003), Basset et al. (2006), Immergluck and Smith (2006), Mallach (2006, 2012), Samsa (2008), Harding et al. (2009), Campbell et al. (2011), Kelly (2013), Silverman et al. (2013), Whitaker and Fitzpatrick (2013), Morckel (2014), Raleigh and Galster (2015), Dewar et al. (2015), Han (2017, 2019), vom Hofe et al. (2019), Wang and Immergluck (2019), Jeon and Kim (2020).

[22] See Lieb et al. (1974), Keenan et al. (1999), Power and Mumford (1999), Bowman and Pagano (2000), Brueckner (2000), Cheshire and Sheppard (2002), Glock and Häußermann (2004), Dixon and Adams (2008), McLaughlin (2012), Newman et al. (2018), Hwang and Lee (2019).

[23] These claims may be directly against built structures (Keenan et al. 1999, Power and Mumford 1999, Lieb et al. 1974), but also on unused and fractal land parcels (Bowman and Pagano 2000; Ho and Spoor 2006; Dixon and Adams 2008; Durst and Ward 2015; Newman et al. 2018). In the case of Milan (see Appendix A), a combination of the two claims occurs.

and Häußermann 2004; Keenan et al. 1999; Newman et al. 2018; Cheshire and Sheppard 2002).

4. *Equity reasons.*[24] In this case, abandonment is considered undesirable because it seems unjust to have abandoned buildings in front of so many people who have problems in entering housing markets (e.g. homeless, migrants, youngsters). Many stress how abandoned buildings exacerbate problems of unequal allocation of resources. Claims are here based on principles of allocative justice in order to find the best accommodation for all urban users. In this view, many suggest regulating urban markets, for instance by increasing property taxes for unused buildings to expand the available offer of spaces.[25]

3.4 Policies for Coping with Abandonment

This sub-section proposes a classification of various forms of interventions that are qualitatively listed according to the level/intensity of *prescriptiveness* and *abstractness* of the measures for private agents (e.g. high, medium, low). The classification mainly accounts for policies implemented at the local level, highlighting the difference between general and specific targets (i.e. category-based, agent-based, area-based).[26] Let us consider them in detail.

1. *Normative policies.* In the case of normative policies, public authorities typically set pragmatic duties through category-based abstract norms imposing commissions or prohibiting certain actions for private agents. For instance, there are positive norms (e.g. requiring maintenance standards for every building) and negative norms, which ask private agents to omit certain actions (e.g. always avoid nuisances).[27] Here the problem of abandonment is considered in general terms: it may emerge always and anywhere in a city.

2. *Enforcing policies.* In the case of enforcing policies, public authorities compel owners to comply with their pragmatic duties through agent-based measures. In this regard, authorities typically use ordinances to directly oblige a single owner to intervene or even expropriate his/her property.[28] Here abandonment is connected with particular problematic situations: for instance, a specific building creating certain risks and harms in a neighbourhood.

---

[24] See Cheshire and Sheppard (2002), McCarthy (2002), Burton (2003), Wyatt (2008), Hirokawa and Gonzalez (2010), Dewar and Weber (2012), Henderson (2015), Gentili and Hoekstra (2018).

[25] For some examples in the case of residential buildings/dwellings, see Wyatt (2008), Hirokawa and Gonzalez (2010) and Henderson (2015).

[26] A "category" here is a class of an indefinite number of non-assignable individuals/agents (e.g. "drivers", "pedestrians", "shopkeepers", "homeowners"). See Moroni (2018).

[27] Building codes do not typically account for negative externalities, but mainly nuisances (Accordino and Johnson 2000, p. 313).

[28] Possible limitations of this approach are on the economic sustainability related to all costs in place which are particularly time-consuming for private entities (e.g. trials). Issues such as liability or just compensation for the owners remain crucial in this process (Kraut 1999, p. 1161).

3.   *Enabling policies.* In the case of enabling policies, public authorities stimulate
       the fulfilment of pragmatic duties through category-based abstract norms. Here,
       the general idea is that abandonment can be avoided/prevented by fostering
       initiatives for building upgrades. This occurs through fiscal incentives such as
       tax abatements or direct subsidies. Another strategy is to simplify or eliminate
       certain unnecessary regulations in order to stimulate redevelopment (Taylor
       et al. 1992; Gu et al. 2019; Moroni and Lorini 2017; Olivadese et al. 2017).[29]
4.   *Collaborative policies.* In the case of collaborative policies, public authori-
       ties motivate the fulfilment of pragmatic duties through agent-based specific
       interventions by promoting public–private partnerships or favouring specific
       community alliances. Here the assumption is that abandonment problems may
       be solved by reducing risk perception and targeting initiatives and resources
       (Aiyer et al. 2015; Sampson et al. 2017; see also Thomson 2011; Savini et al.
       2015).
5.   *Infrastructural policies.* In the case of infrastructural policies, public authori-
       ties support the fulfilment of pragmatic duties through site-based and specific
       interventions. For instance, the redevelopment of abandoned buildings or plots
       may be stimulated by investing in public spaces and equipment (Galster 2019;
       Raleigh and Galster 2015; Dewar and Weber 2012; Haase et al. 2014); for
       example, by providing a new metro station, or, more generally, by favouring
       transit-oriented developments (e.g. TODs: Abe et al. 2014; Hong et al. 2016)
       (Table 3.4; Fig. 3.1).

---

[29] We may define regulations as "unnecessary" when they increase private and/or public costs
without any collective advantage (Cozzolino and Moroni 2021).

Table 3.4 Types and characteristics of policies

Types of policies	Degree of prescriptiveness	Degree of abstraction	Target	Relationship with pragmatic duties
Normative e.g. maintenance requirements, avoid nuisances	Medium	High	Category-based (e.g. all private owners in a city)	*Set* pragmatic duties
Enforcing e.g. ordinances of expropriation, demolition, rehabilitation	High	Low	Agent-based (e.g. the ownership)	*Compel* compliance with pragmatic duties
Enabling e.g. through subsidies, simplification of bureaucracy	Low	High	Category-based (e.g. private sector)	*Stimulate* the fulfilment of pragmatic duties
Collaborative e.g. PPPs: public–private partnerships, CIDs: common interest developments	Low	Medium	Agent-based (e.g. specific public, private, non-profit actors)	*Motivate* the fulfilment of pragmatic duties
Infrastructural e.g. TODs: transit-oriented developments, new infrastructures and public amenities	/	Low	Site-based (e.g. specific areas in a city)	*Support* the fulfilment of pragmatic duties

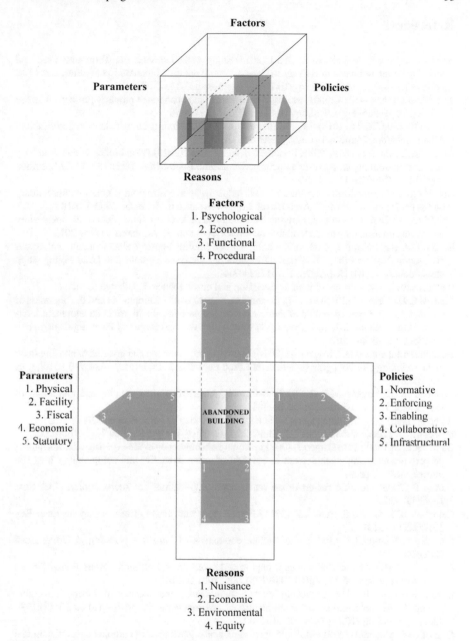

Fig. 3.1 General categories in the analytical schema

References

Abe S, Nakagawa D, Matsunaka R, Oba T (2014) Study on the factors to transform underused land focusing on the influence of railway stations in central areas of Japanese Local cities. Land Use Policy 41:344–356. https://doi.org/10.1016/j.landusepol.2014.06.021

Accordino J, Johnson GT (2000) Addressing the vacant and abandoned property problem. J Urban Aff 22(3):301–315. https://doi.org/10.1111/0735-2166.00058

Adobati F, Garda E (eds) (2018) Biografie sospese. Un'esplorazione dei luoghi densamente disabitati della Lombardia. Mimesis/Kosmos, Sesto San Giovanni

Aiyer SM, Zimmerman MA, Morrel-Samuels S, Reischl TM (2015) From broken windows to busy streets: a community empowerment perspective. Health Educ Behav 42(2):137–147. https://doi.org/10.1177/1090198114558590

ANCE (2012) Osservatorio congiunturale sull'industria delle costruzioni [Economic observatory on the construction industry]. Available at: http://www.ance.it. Accessed 28 Oct 2018

ANCE (2019) Osservatorio congiunturale sull'industria delle costruzioni [Economic observatory on the construction industry]. Available at: https://www.ance.it. Accessed 15 Nov 2019

Basset EM, Schweitzer J, Panken S (2006) Understanding housing abandonment and owner decision-making in Flint, Michigan: an exploratory analysis. Lincoln Inst Land Policy (sep). https://doi.org/10.1017/CBO9781107415324.004

Baum A (1991) Property investment depreciation and obsolescence. Routledge, London

Bennett L, Dickinson J (2015) Forcing the empties back to work? Ruinphobia and the bluntness of law and policy. Paper presented at the transience and permanence in urban development international research workshop, University of Sheffield, Town and Regional Planning Department, Sheffield, 14–15 Jan 2015

Bogataj D, McDonnell DR, Bogataj M (2016) Management, financing and taxation of housing stock in the shrinking cities of aging societies. Int J Prod Econ 181:2–13. https://doi.org/10.1016/j.ijpe.2016.08.017

Bowman AOM, Pagano MA (2000) Transforming America's cities: policies and conditions of vacant land. Urban Aff Rev 35(4):559–581

Brand S (1995) How buildings learn: what happens after they're built. Penguin, New York

Brueckner JK (2000) Urban sprawl: diagnosis and remedies. Int Reg Sci Rev 23(2):160–171

Buitelaar E, Moroni S, De Franco A (2021) Building obsolescence in the evolving city. Reframing property vacancy and abandonment in the light of urban dynamics and complexity. Cities 108 (online ahead of print)

Burton E (2003) Housing for an urban renaissance: implications for social equity. Hous Stud 18(4):537–562

Campbell BJY, Giglio S, Pathak P (2011) Forced sales and house prices. American Econ Rev 101(5):2108–2131

Cheshire P, Sheppard S (2002) The welfare economics of land use planning. J Urban Econ 52(2):242–269

Cohen JR (2001) Abandoned housing: exploring lessons from Baltimore. Hous Policy Debate 12(3):415–448. https://doi.org/10.1080/10511482.2001.9521413

Cozens P, Tarca M (2016) Exploring housing maintenance and vacancy in Western Australia: perceptions of crime and crime prevention through environmental design. Prop Manag 34(3):199–220. https://doi.org/10.1108/PM-06-2015-0027

Cozzolino S, Moroni S (2021) Multiple agents and self-organisation in complex cities: the crucial role of several property. Land Use Policy 103:1–7

Dewar M, Weber MD (2012) City abandonment. In: Dewar M, Weber MD (eds) The Oxford handbook of urban planning. Oxford University Press, Oxford, pp 563–586

Dewar M, Seymour E, Druță O (2015) Disinvesting in the city: the role of tax foreclosure in Detroit. Urban Aff Rev 51(5):587–615

Dixon T, Adams D (2008) Housing supply and brownfield regeneration in a post-Barker world: is there enough brownfield land in England and Scotland? Urban Studies 45(1):115–139

Durst NJ, Ward PM (2015) Lot vacancy and property abandonment: colonias and informal subdivisions in Texas. Int J Hous Policy 15(4):377–399. https://doi.org/10.1080/14616718.2015.109 0095

Edson CL (1972) Housing abandonment—the problem and a proposed solution. Real Property, Probate Trust J 7(2):382–390

Ellickson RC (1996) Controlling chronic misconduct in city spaces: of panhandlers, skid rows, and public-space zoning. Yale Law J 105(5):1166–1248

Farris JT (2001) The barriers to using urban infill development to achieve smart growth. Hous Policy Debate 12(1):41–45. https://doi.org/10.1080/10511482.2001.9521397

Foster A, Newell JP (2019) Detroit's lines of desire: footpaths and vacant land in the Motor City. Landsc Urban Plan 189(April):260–273. https://doi.org/10.1016/j.landurbplan.2019.04.009

Fuentes JM, Gallego E, García AI, Ayuga F (2010) New uses for old traditional farm buildings: the case of the underground wine cellars in Spain. Land Use Policy 27(3):738–748

Galster G (2001) On the nature of neighbourhood. Urban Studies 38(12):2111–2124

Galster G (2019) Why shrinking cities are not mirror images of growing cities: a research agenda of six testable propositions. Urban Aff Rev 55(1):355–372. https://doi.org/10.1177/107808741 7720543

Garda E (2018) Negli spazi vuoti della metropoli: esperienze di riuso collettivo tra temporaneità e permanenze. Geogr Notebooks 1(2):97–110

Gentili M, Hoekstra J (2018) Houses without people and people without houses: a cultural and institutional exploration of an Italian paradox. Hous Stud 34(3):425–447. https://doi.org/10.1080/02673037.2018.1447093

Glaeser EL, Gyourko J (2005) Urban decline and durable housing. J Polit Econ 113(2):345–375. https://doi.org/10.1086/427465

Glock B, Häußermann H (2004) New Trends in urban development and public policy in eastern Germany: dealing with the vacant housing problem at the local level. Int J Urban Reg Res 28(December):919–929

Grover R, Grover C (2015) Obsolescence—a cause for concern?. J Property Invest Fin

Gu D, Newman G, Kim JH, Park Y, Lee J (2019) Neighborhood decline and mixed land uses: mitigating housing abandonment in shrinking cities. Land Use Policy 83(February):505–511. https://doi.org/10.1016/j.landusepol.2019.02.033

Haase A, Rink D, Grossmann K, Bernt M, Mykhnenko V (2014) Conceptualizing urban shrinkage. Environ Plan A 46:1519–1534. https://doi.org/10.1068/a46269

Han HS (2017) Neighborhood characteristics and resistance to the impacts of housing abandonment. J Urban Aff 39(6):833–856

Han HS (2019) Exploring threshold effects in the impact of housing abandonment on nearby property values. Urban Aff Rev 55(3):772–799. https://doi.org/10.1177/1078087417720303

Harding JP, Rosenblatt E, Yao VW (2009) The contagion effect of foreclosed properties. J Urban Econ 66(3):164–178. https://doi.org/10.1016/j.jue.2009.07.003

Heller MA (1998) The tragedy of the anticommons: property in the transition. Harv Law Rev 111(3):621–688

Henderson SR (2015) State intervention in vacant residential properties: an evaluation of empty dwelling management orders in England. Eviron Plann C Gov Policy 33(1):61–82. https://doi.org/10.1068/c12215

Hillier AE, Culhane DP, Smith TE, Tomlin CD (2003) Predicting housing abandonment with the Philadelphia neighborhood information system. J Urban Aff 25(1):91–106. https://doi.org/10.1111/1467-9906.00007

Hirokawa KH, Gonzalez I (2010) Regulating vacant property. The Urban Lawyer 42(3):627–637. https://www.jstor.org/stable/27895816

Ho P, Spoor M (2006) Whose land? The political economy of land titling in transitional economies. Land Use Policy 23(4):580–587. https://doi.org/10.1016/j.landusepol.2005.05.007

Holtslag-Broekhof S (2018) Urban land readjustment: necessary for effective urban renewal? Analysing the Dutch quest for new legislation. Land Use Policy 77(August):821–828. https://doi.org/10.1016/j.landusepol.2017.07.062

Hong S, Shin E, Kim S (2016) Effects of frequently overriding regulations on urban renewal in Seoul: a warning. J Urban Plann Dev 142(3). https://doi.org/10.1061/(ASCE)UP.1943-5444.0000311

Hughes C, Jackson C (2015) Death of the high street: identification, prevention, reinvention. Reg Stud Reg Sci 2(1):237–256. https://doi.org/10.1080/21681376.2015.1016098

Humphris I, Rauws W (2020) Edgelands of practice: post-industrial landscapes and the conditions of informal spatial appropriation. Landscape Research (online ahead of print)

Hwang SW, Lee SJ (2019) Unused, underused, and misused: an examination of theories on urban void spaces. Urban Res Pract 13(5):1–17. https://doi.org/10.1080/17535069.2019.1634140

Immergluck D, Smith G (2006) The impact of single-family mortgage foreclosures on neighborhood crime. Hous Stud 21(6):851–866. https://doi.org/10.1080/02673030600917743

ISTAT (2014) Edifici e abitazioni [Buildings and homes]. https://www.istat.it. Accessed 21 Feb 2019

Jacobs J (1961) The death and life of great American cities. Vintage, New York

Jacquette D (2002) Ontology. Acumen, Chesham

Jennings CR (1999) Socioeconomic characteristics and their relationship to fire incidence: a review of the literature. Fire Technol 35(1):7–34. https://doi.org/10.1023/A:1015330931387

Jeon Y, Kim S (2020) Housing abandonment in shrinking cities of East Asia: case study in Incheon, South Korea. Urban Studies 57(8):1749–1767. https://doi.org/10.1177/0042098019852024

Keenan P, Lowe S, Spencer S (1999) Housing abandonment in inner cities-the politics of low demand for housing. Hous Stud 14(5):703–716. https://doi.org/10.1080/02673039982687

Kelly JJ (2013) A continuum in remedies: reconnecting vacant houses to the market. Saint Louis Univ Public Law Rev 33(3):109–139

Kraut DT (1999) Hanging out the no vacancy sign: eliminating the blight of vacant buildings from urban areas. New York Univ Law Rev 74(4):1139–1177

Lami IM (eds) (2020) Abandoned buildings in contemporary cities: smart conditions for actions. In: Smart innovation, systems and technologies, vol 168. Springer, Cham. https://doi.org/10.1007/978-3-030-35550-0_9

Lieb RC, Merel RA, Perlin AS, Sadoff MB (1974) Abandonment of residential property in an urban context. DePaul Law Rev 23(3):1186–1224

Mallach A (2006) Bringing buildings back: from abandoned properties to community assets. Rutgers University Press, New Brunswick (NJ)

Mallach A (2012) Depopulation, market collapse and property abandonment: surplus land and buildings in legacy cities. In: Mallach A (ed) Rebuilding America's legacy cities: strategies for cities losing population. American Assembly, New York, pp 85–110

Marini S, Santangelo V (eds) (2013) New life cycles for architecture and infrastructure of city and landscape. Aracne, Roma

McCarthy L (2002) The brownfield dual land-use policy challenge. Land Use Policy 19:287–296. https://doi.org/10.1016/S0264-8377(02)00023-6

McLaughlin RB (2012) Land use regulation: where have we been, where are we going? Cities 29:S50–S55

Merrett S, Smith R (1986) Stock and flow in the analysis of vacant residential property. Town Plann Rev 57(1):51–67

Micelli E, Pellegrini P (2019) Paradoxes of the Italian historic centres between underutilisation and planning policies for sustainability. Sustainability 11(9):2614. https://doi.org/10.3390/su11092614

Molloy R (2016) Long-term vacant housing in the United States. Reg Sci Urban Econ 59:118–129

Morckel V (2014) Predicting abandoned housing: does the operational definition of abandonment matter? Community Dev 45(2):121–133. https://doi.org/10.1080/15575330.2014.892019

Morgan DJ (1980) Residential housing abandonment in the United States: the effects on those who remain. Environ Plan A 12(12):1343–1356. https://doi.org/10.1068/a121343

Moroni S (2018) Public Interest. In: Gunder M, Madanipour A, Watson V (eds) The Routledge handbook of planning theory. Routledge, London, pp 69–80

Moroni S, Lorini G (2017) Graphic rules in planning: a critical exploration of normative drawings starting from zoning maps and form-based codes. Plan Theory 16(3):318–338

Moroni S, De Franco A, Bellè BM (2020a) Vacant buildings. Distinguishing heterogeneous cases: public items versus private items; empty properties versus abandoned properties. In: Lami IM (ed) Abandoned buildings in contemporary cities: smart conditions for actions. Springer, Cham, pp 9–18

Moroni S, De Franco A, Bellè BM (2020b) Unused private and public buildings: re-discussing merely empty and truly abandoned situations, with particular reference to the case of Italy and the city of Milan. J Urban Aff 42(8):1299–1320. https://doi.org/10.1080/07352166.2020.179 2310

Németh J, Langhorst J (2014) Rethinking urban transformation: temporary uses for vacant land. Cities 40:143–150. https://doi.org/10.1016/j.cities.2013.04.007

Newman G, Park Y, Lee RJ (2018) Vacant urban areas: causes and interconnected factors. Cities 72:421–429

Olivadese R, Remøy H, Berizzi C, Hobma F (2017) Reuse into housing: Italian and Dutch regulatory effects. Prop Manag 35(2):165–180. https://doi.org/10.1108/PM-10-2015-0054

Popper K (1974) Unended quest: an intellectual autobiography. Open Court, Chicago

Power A, Mumford K (1999) The slow death of great cities? Urban abandonment or urban renaissance. York Publishing Services Ltd., York

PWC (PricewaterhouseCoopers) (2020) The Italian NPL market. Ready to face the crisis. Available at: https://www.pwc.com. Accessed 24 Nov 2020

Raleigh E, Galster G (2015) Neighborhood disinvestment, abandonment, and crime dynamics. J Urban Aff 37(4):367–396

Remøy H (2010) Out of office. A study on the cause of office vacancy and transformation as a means to cope and prevent. IOS Press, Delft

Remøy H, Street E (2018) The dynamics of 'post-crisis' spatial planning: a comparative study of office conversion policies in England and the Netherlands. Land Use Policy 77:811–820

Ross CE, Mirowsky J (1999) Disorder and decay: the concept and measurement of perceived neighborhood disorder. Urban Aff Rev 34(3):412–432. https://doi.org/10.1177/107808749903 400304

Ross GM, Portugali J (2018) Urban regulatory focus: a new concept linking city size to human behaviour. R Soc Open Sci 5(5):1–11. https://doi.org/10.1098/rsos.171478

Sampson N, Nassauer J, Schulz A, Hurd K, Dorman C, Ligon K (2017) Landscape care of urban vacant properties and implications for health and safety: lessons from photovoice. Health Place 46(May):219–228. https://doi.org/10.1016/j.healthplace.2017.05.017

Samsa MJ (2008) Reclaiming abandoned properties: using public nuisance suites and land banks to pursue economic redevelopment. Cleveland State Law Rev 56(1):189–232

Savini F, Majoor S, Salet W (2015) Dilemmas of planning: intervention, regulation, and investment. Plan Theory 14(3):296–315

Schilling J (2009) Code enforcement and community stabilization: the forgotten first responders to vacant and foreclosed homes. Albany Govern Law Rev 2:101–162

Searle J (1995) The construction of social reality. The Free Press, New York. La costruzione della realtà sociale. Einaudi, Torino, 2006

Seo W, von Rabenau B (2011) Spatial impacts of microneighborhood physical disorder on property resale values in Columbus, Ohio. J Urban Plann Dev 137(3):337–345. https://doi.org/10.1061/ (ASCE)UP.1943-5444.0000067

Silverman RM, Yin L, Patterson KL (2013) Dawn of the dead city: an exploratory analysis of vacant addresses in Buffalo, NY 2008–2010. J Urban Aff 35(2):131–152. https://doi.org/10.1111/j.1467-9906.2012.00627.x

Spelman W (1993) Abandoned buildings: magnets for crime? J Crim Just 21(5):481–495. https://doi.org/10.1016/0047-2352(93)90033-J

Sternlieb G, Burchell RW, Hughes JW, James FJ (1974) Housing abandonment in the urban core. J Am Plann Assoc 40(5):321–332. https://doi.org/10.1080/01944367408977488

Talen E (2012) Zoning and diversity in historical perspective. J Plan Hist 11(4):330–347. https://doi.org/10.1177/1538513212444566

Taylor RB, Koons BA, Kurtz EM, Greene JR, Perkins DD (1992) Street blocks with more nonresidential land use have more physical deterioration. Urban Aff Rev 31(1):120–136

Thomsen A, Van der Flier K (2011) Obsolescence and the end of life phase of buildings. In: Proceedings of the CIB international conference management and innovation for a sustainable built environment—MISBE2011, Amsterdam, The Netherlands, 20–23 June 2011

Thomson DE (2011) Strategic geographic targeting in community development: examining the congruence of political, institutional, and technical factors. Urban Aff Rev 47(4):564–594

vom Hofe R, Parent O, Grabill M (2019) What to do with vacant and abandoned residential structures? The effects of teardowns and rehabilitations on nearby properties. J Reg Sci 59(2):228–249. https://doi.org/10.1111/jors.12413

Wallace D, Schalliol D (2015) Testing the temporal nature of social disorder through abandoned buildings and interstitial spaces. Soc Sci Res 54:177–194

Wang K, Immergluck D (2019) Housing vacancy and urban growth: explaining changes in long-term vacancy after the US foreclosure crisis. J Housing Built Environ 34(2):511–532. https://doi.org/10.1007/s10901-018-9636-z

Wegmann J (2020) Residences without residents: assessing the geography of ghost dwellings in big US cities. J Urban Aff 42(8):1103–1124

Whitaker S, Fitzpatrick TJ (2013) Deconstructing distressed-property spillovers: the effects of vacant, tax-delinquent, and foreclosed properties in housing submarkets. J Hous Econ 22(2):79–91. https://doi.org/10.1016/j.jhe.2013.04.001

White MJ (1986) Property taxes and urban housing abandonment. J Urban Econ 20(3):312–330. https://doi.org/10.1016/0094-1190(86)90022-7

WWF (2013) Riutilizziamo l'Italia. Dal censimento del dismesso scaturisce un patrimonio di idee per il futuro del Belpaese, Edizioni WWF. Available at: https://wwfit.awsassets.panda.org. Accessed 3 April 2018

Wyatt P (2008) Empty dwellings: the use of council-tax records in identifying and monitoring vacant private housing in England. Environ Plan A 40(5):1171–1184. https://doi.org/10.1068/a39176

Chapter 4
Discussion

Abstract This section is divided into two parts. The first highlights the difference between agentive and non-agentive functions (Sect. 4.1). The difference between having status and having responsibilities deriving from that status is considered. The second part employs our analytical schema in the light of empirical investigations (Sect. 4.2). It discusses the role of parameters, the multiplicity of factors, suggesting a plurality of policies, while critically revisiting the reasons that qualify abandonment as a negative phenomenon in cities.

Keywords Agentive function · Non-agentive functions · Multiplicity · Plurality

4.1 Agentive and Non-agentive Functions

Entities can have *non-agentive* and *agentive* functions (Searle 1995, p. 20 ff.). Non-agentive functions do not depend on human agents; they exist in the physical world independently from us. Instead, agentive functions do depend on us; their existence and persistence are human-dependent.

In regard to agentive functions, Searle (1995, pp. 20–21) stresses:

> Sometimes the assignment of function has to do with our immediate purposes, whether practical, gastronomic, aesthetic, educational or whatever. When we say, 'This stone is a paperweight', 'This object is a screwdriver' or 'This is a chair', these three functional notions mark *uses* to which we put objects, functions that we do not discover and that do not occur naturally, but that are assigned relative to the practical interests of conscious agents. … Thus bathtubs, coins and screwdrivers *require continued use* on our part in order to function as bathtubs, coins and screwdrivers, but heart and livers continue to function as hearts and livers even when no one is paying any attention (emphasis added).[1]

In the light of all this and resuming to our main issue, we may observe that buildings matter to us mainly for their agentive functions, rather than for their non-agentive functions.

---

[1] Further elaborations on Searle's distinction between agentive and non-agentive functions can be found in Kroes (2003), Tieffenbach (2010) and Loddo (2016). Loddo (2016) notes that agentive functions are *introduced* by social agents and allow them to adapt/modify reality, while non-agentive functions are *discovered* by social agents and represent mostly constraints and uncertainties of reality for them.

A. De Franco, *Abandonment as a Social Fact*, SpringerBriefs in Geography,
https://doi.org/10.1007/978-3-030-90367-1_4

Table 4.1 Relations between ownership status and responsibilities

		Status (i.e. ownership)	
		Yes	No
Responsibility (e.g. upkeep)	Yes	A Responsible owner	B Guardians, trustees, managers, etc.
	No	C Neglectful owner	D Everyone else

Abandoned buildings, as such, lose certain agentive functions (e.g. habitability) when owners do not fulfil their duties (i.e. do not maintain their assets). In this sense, (abandoned) buildings progressively return to their "brute" nature.

This is interestingly connected with the distinction introduced in Chap. 2 between *animus derelinquendi* (i.e. the abandoning subject's intentions) and *res derelictae* (i.e. the abandoned object). The latter does not have agentive functions in itself; agentive functions have been lost because of the former.

In conclusion, "abandonment" as a social fact is linked to a loss of agentive functions.

Additional clarification on abandonment and agentive functions might be added now: that is the difference between (i) having a *status* and (ii) having the *responsibilities* deriving from that status.[2]

Table 4.1 attempts to summarize how the abandonment problem is related to the different roles and duties performed by social agents. As regards owners, the problem of abandonment depends on whether the agent fulfils the duties attached to the ownership title (cells A and C). Obviously, an agent can take responsibility even without being an owner (cell B); from a logical point of view, there are also other agents who have no responsibility or status with respect to certain assets (cell D).

To understand the abandonment issue, we need to relate it to an institutional "invisible ontology" that establishes relations between physical objects and social agents (Searle 1995). Thus, the owners but also guardians, etc., have relational sets of *pragmatic duties* that are assigned or acquired by formal declarations (e.g. property acts, contracts).[3] Moreover, there are other *absolute duties* that must be respected by "everyone else", for instance, do not enter or damage someone else's property.

---

[2] The same difference may be traced between the possibility of renouncing *a status* and the possibility of renouncing *the responsibilities* deriving from that status. This study sustains that it is not possible to renounce to such responsibilities, while it does not deal directly with the possibility of renouncing ownership status (on this, see Romano 2002; Peñalver 2010).

[3] In the cases here illustrated the focus has been on "formal" aspects of property titles and duties (on informal situations, see anyway Humpris and Rauws 2020).

4.2 Back to the Schema

4.2.1 Parameters

Both as regards *parameters* and *indicators* there are various definitional and operative issues to consider. To expressly recognize certain buildings as "abandoned"
means to identify abandonment as a social fact, not as a brute one. Even when we
employ physical indicators to identify abandonment, *they are merely signs/signals
of social facts*. For instance, a falling eave of a house may be indicative of a state
of neglect of a building but does not demonstrate in itself abandonment (e.g. the
building may be occupied). At the same time, a missing roof can be indicative of
a state of abandonment, not only as a product of brute physical forces (e.g. time,
weathering) but as an effect of a rule that discourages property maintenance.[4]

4.2.2 Factors

As regards abandonment *factors*, this study stresses that (i) they can be understood
only as *conditions* and not as *causes* and (ii) they act *cumulatively* and *simultaneously*
without the possibility of attributing to any of them an explanatory priority that is
always valid.

In the first case the point is that there is no linear and deterministic cause-effect
relation in abandonment processes, but some background conditions that can make
some outcomes more probable.[5] Thus, factors can be regarded as multipliers or trig-
gers that influence behaviours (including investment attitudes, agents' interactions,
etc.).[6] This also depends on the fact that we are not just talking about brute facts but
about social facts (social entities and processes).

In the second case, the point is that it is not possible to attribute the phenomenon
of abandonment to a single aspect—as some tend to do, for example, attributing it
almost exclusively to the financialization of the real estate market[7]—but always to a

---

[4] In Italy, for instance, Confedilizia (2019) pointed out that the special tax-discount regime for
ruined buildings (i.e. *unità collabenti*: D.M. 28 of 1998, art. 3, comma 2; Dpr. 445 of 2000, D.M. 28
of 1998, D.Lgs. 504 of 1992) is influencing the spread of an "unwilling attitude" of Italian owners
to maintain their assets.

[5] See on this also Grossmann and Haase (2016), Galster (2019), Gu et al. (2019, p. 509), Han (2019).

[6] On this see Basset et al. (2006), Schilling (2009, p. 132), Németh and Langhorst (2014, p. 149),
Aiyer et al. (2015), Galster (2019, p. 366).

[7] See Aalbers (2016), Wang and Immergluck (2019), Atkinson (2019), Wegmann (2020). Based on
Milan's case (see Appendix A), it is clear that many of the buildings were abandoned long before the
2008's economic crisis. Thus, most of the buildings are visibly old sheds and warehouses inherited
by the industrial past of the city. Some of the assets (mostly small-sized properties) were redeveloped
promptly after the first warrants issued by the municipality, presumably, by physical persons (see
Appendix B for details). This resizes the hypothesis of direct linkages between abandonment and
financialization interests pushed by large investment groups, at least at present, in Milan.

set of elements that combine and sometimes reinforce each other.[8] For example, if the economic crisis may have exacerbated the problem and some public rules make it difficult to transform certain assets, these two factors add up and in turn have repercussions on the attitudes of the actors involved.[9]

Heavy bureaucracy and property taxes affect the accumulation of abandoned buildings in various ways. As Resta (2018, pp. 258–259) observes, some critical passages recently emerging in administrative jurisprudence tend to limit the owners' discretion (e.g. on the custody of an unproductive asset) to avoid community burdens (e.g. externalities). Although, before restricting the agents' span of choices, it seems legitimate to ask whether public rules can be readjusted to prevent/anticipate certain problems before they escalate. Even when policymakers consider the abandonment of private buildings as a collective problem, one cannot overlook the need to reform certain public rules and procedures (e.g. legislative and bureaucratic simplification; see Moroni 2015b, Olivadese et al. 2017; OPPAL 2019) and forms of taxation.[10]

The cumulative effect and simultaneity of psychological, procedural and functional factors in many cases shows that any reductionism should be avoided and that it is not possible to make quantitative specific explanations and predictions of abandonment; for instance, *only* in economic terms. Explanations involving certain predefined sequences of stages or phases risk falling into this mistake (Schilling 2009, pp. 121–125) (Fig. 4.1).[11]

Against the idea that all estate markets have been heavily financialized and the corollary hypothesis that many buildings are unused just for this reason, some data and figures should be considered. In Italy, for instance, the percentage of building

---

[8] By contrast, Lieb et al. (1974, p. 1193) state: "the causes precipitating the negative reinvestment attitude experienced by [landlords, tenants, financial institutions, government and public agencies] are for the most part mutually exclusive of one another".

[9] In the literature many authors deployed physiological metaphors to express the gravity of abandonment processes (i.e. contagion effects: Greenberg et al. 1990; Hillier et al. 2003; Galster 2019; Han 2019). These metaphors often have a strong evocative power influencing the idea that local contexts need direct and large-scale programs of interventions (Hillier et al. 2003; Silverman et al. 2013).

[10] Considering the long-standing relation between property taxation and the abandonment problem (Dewar et al. 2015), a possible tool is "land-value taxation" as also suggested by Accordino and Johnson (2000), Raslanas et al. (2010), Vincent (2012), Cho et al. (2013). In Italy this would require a structural fiscal reform (possibly at the national level; Moroni and Minola 2019), which is also what the European Commission long ago recommended in order to make tax expenditures more efficient. In 2014, the government had advanced the hypothesis of redistributing the tax burden between urban centres and suburbs with a land registry reform. According to Confedilizia (2019, 2015), this prospect could conceal yet another manoeuvre by the Italian state to have more revenue, without resolving the discrepancies between the values recorded by public agencies with the profile of census areas. In this regard, the *Corte di Cassazione* expressed itself in 2018 noting that the adjustment of the sale price of a property cannot be justified based solely on present fiscal data, underlining that there is no perfect overlap between cadastral value and market or "normal" value of the sale (i.e. deviations of real estate values from OMI data; see the sentence 12,269/2018, 18 May).

[11] On the idea of stages/phases, see Lieb et al. (1974), Sternlieb et al. (1974), Wilson et al. (1994), Long et al. (2011).

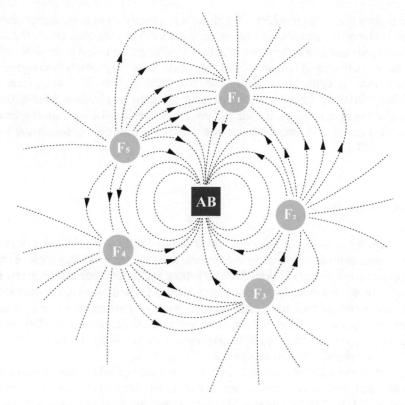

Fig. 4.1 Multiple factors (F1 … Fn) conditioning abandonment

units owned by physical persons is 88.63% (57,087,773 units), while that of non-physical persons is 11.37% (7,323,276). Moreover, the percentage of building units at the owners' disposal (i.e. not used continuously and therefore comprising also vacant and abandoned buildings) is for physical persons 99.36% (6,265,385) and for non-physical persons 0.64% (40,175).[12]

Even in a metropolitan city like Milan, the number of dwellings owned by physical persons is large (632,361 units): 66.1% is used as the primary residence, 19.2% is rented and 8.3% is "unused" (52,321 units).[13]

The case of Milan also demonstrates the impossibility of identifying a single cause responsible for the phenomenon of abandonment and that there is rather a plurality of simultaneous conditions. In contrast, certain local policy makers assume that speculation of real estate funds is the only real motive for keeping certain assets abandoned in the city, while certain regional policymakers assume that an economic crisis is the

---

[12] Elaboration based on MEF data. See https://www1.finanze.gov.it/finanze3/immobili/#/tabelle (Accessed February 2021). See Appendix B for details.

[13] Unfortunately, the information for non-physical persons is not available at the subnational scale. Estimates from MEF (2019), 2016 data.

main explanation of the problem. This attitude to identify a single main explanatory factor is one of the reasons for clashes between local and regional levels.[14] But if there was only one factor, the set of abandoned buildings identified by local planning documents (Comune di Milano 2019a, b) would have very similar/homogeneous characteristics in terms of ownership, location, building type, etc., while there are significant differences among them. For instance, there are abandoned buildings in central and peripheral areas, in rich and poor areas, residential and non-residential units, all of the large and small dimensions,[15] and with different typologies and years of construction.[16]

4.2.3 Reasons

As regards the *reasons* according to which abandonment is considered a negative phenomenon in urban contexts, we can observe that those who advance some of them tend to see abandonment as a brute fact and to underestimate its social dimension. For example, those who consider abandonment a problem for reasons of environmental protection or equity treat some urban aspects mainly as quantitative and allocative issues, as if they were just brute facts: in the first case, the abandoned building owned by X and the open land owned by Y; in the second case, the abandoned building owned by X and the absence of a house/space for Z (Fig. 4.2).

It is interesting for instance to note that, in the case of Milan, the identification of abandoned buildings takes place within the "Land Consumption Chart" (Comune di Milano 2019b). The background idea seems that land transformation issues are a problem of "communicating vessels". In this perspective, the abandoned buildings owned by certain subjects and the open lands owned by others, are treated at the same level, as homogeneous interchangeable positions. As a consequence, everything is seen as a problem of (dis)equilibrium and waste in a rational-comprehensive view. On the contrary, the elements at stake are not placed on the same plan—nor they are

---

[14] Some notion of causality is necessary, but this is a complex matter in human science (Smith 2020, p. 24). By contrast, also in the Milan case, policy makers often opt for single—and sometimes antagonistic—causes. See the debate at https://it.businessinsider.com/milano-attacca-la-regione-sugli-edifici-abbandonati-bonus-per-non-ristrutturare-e-un-regalo-agli-immobiliaristi/ (Accessed May 2020).

[15] A rough estimate of the dimension of the building stock (expressed in gross surface areas) can be subdivided as follows: 38% of the properties are below 1000 m^2 (65 units); 47% between 1001 and 10,000 m^2 (80 units); 14% between 10,001 and 100,000 m^2 (23 units); only one property is more than 100,000 m^2.

[16] A general classification of the main building typologies can be summarised as follows: 35% of the buildings are *slabs* (i.e. mostly hangars or buildings within 4 floors, 59 units); 24% of the buildings are *linear blocks* (of which 28 units regular longitudinal buildings and 13 L-shaped); 23% of the properties are *compounds* (of which 13 units host mainly linear blocks and 26 slabs); 12% units are open or closed *courts* (20 units); 2% *detached structures* (2 villas and 1 tower). The remaining 4% of the sample (7 units) are classified as "other" since they are not exclusively buildings but also grey fields and logistic areas (e.g. underground parking).

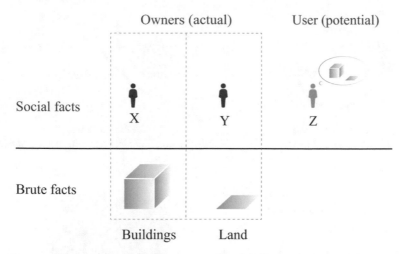

Fig. 4.2 Ontological strata of urban reality

directly comparable—because they embed different titles and rights that pertain to a different *stratum* of reality. Taking this seriously, the metaphor of communicating vessels is subverted by "institutional facts" and rather than a linear equilibrating relation between vessels containing the same liquid, we have a sort of "water and oil situation" (Fig. 4.3).

In this perspective, the too narrow "linear thinking" of certain land use regulations based on environmental claims is open to criticism (McLaughlin 2012). Certain public discourses typically emphasise brute facts while neglecting and even demonising certain social facts (incurring into a typical "whole-society" theory error: Smith 2020, p. 23). This is detrimental both at a conceptual level as a certain rational-comprehensive approach to the issue of scarce resources assumes a "finite" or "complete" world[17]; and at a practical level when excessively limiting urbanization create an inelastic land supply. It is true that we have to be concerned for the environment. However, environmental claims in land-use regulation often establish—and presuppose—causality and equilibrium between two different realms: ecosystems, which are made of "brute" elements and urban systems, mainly made of "social" elements (Marshall 2012).[18]

---

[17] I do not suggest that resources are "unlimited" or "infinite" in a literal sense; anyway, purposeful actions are able to create *new* resources in unexpected ways (e.g. innovation; Kirzner 1997). Note that recycling approaches for urban/environmental waste (as Lynch 1990) cannot be resolutive, since they may underlie positivistic, technocratic and *naif* approaches to describe "waste". Think for instance of the paradox of assigning a high value to passive and depreciated resources for the sole purpose of implementing models of circular economics (Resta 2018, pp. 267–269).

[18] Consider, for instance, another point of view on negative environmental consequences. Abandoned buildings may, after a certain amount of time, facilitate biodiversity in urban areas as plants and animals move in and establish their home. These, somewhat alternative, re-uses of abandoned buildings by non-human entities typically contrast with the interests of their counterparts (e.g. infestations). This, obviously, is not to say that re-naturalisation is bad (actually it is very beneficial for

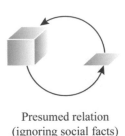 

Presumed relation
(ignoring social facts)

Actual relation
(taking social facts into account, e.g. ownership)

Urban spaces   Natural spaces

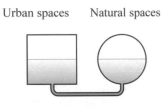

"communicating vessels" view

"water and oil" view

Fig. 4.3 Presumed and actual relations in urban issues

In regard to social equity claims, some authors similarly point out certain "disjunctures" between policy formulation and real-world situations (Henderson 2015). When speaking about the paradox of "many homes without people" and "many people without a home", it is necessary to understand issues as building vacancy in its various nuances (Gentili and Hoekstra 2018). Also, in this case, the fact of having many people struggling to find a place to live is not directly related—or in contrast—to the amount of space available in a city.[19] Issues such as homelessness and poverty may have both absolute and relative meanings, but the most relevant way to tackle both is to work on background *distributive* issues rather than merely ongoing *allocative* ones (Moroni 2020), as we are dealing with a dynamic reality.

Furthermore, when connecting the problems of "abandonment" and that of "poverty" we are actually establishing a link between *two distinct social facts*, not just one.[20]

---

urban living), but to stress our viewpoint on an abandonment problem that is relative and dependent on "social facts".

[19] Additionally, next to negative equity consequences, one may say that abandoned buildings are useful for alternative practices of place making for those who would prefer to remain "outside the radar" (Humpris and Rauws 2020). The suggestion is to wonder what gets lost in the hypothesis that the abandonment issue would be definitively solved in our cities.

[20] Compare with other authors who stress abandonment causes and/or effects in relation to socio-material deprivation (Edson 1972; Sternlieb et al. 1974; Morgan 1980; Merrett and Smith 1986; Wilson et al. 1994; Jennings 1999; Keenan et al. 1999; Kraut 1999; Power and Mumford 1999;

In short, one cannot force empty properties back into use if the protection of the right of someone is at the detriment of the right of someone else. Rights are stronger if these are guaranteed to all individuals indiscriminately; housing units will be more available if urban markets are more elastic and certain rules (e.g. rental contracts) create convenience for all parts at play.[21]

Our way of considering the question is a new formulation of the critique (Rauws 2017) addressed to those who tend to see the city as an "efficientable" simple/closed system rather than a complex/open one.[22]

By contrast, the other two reasons linked to nuisances and economic effects seem more coherent and compatible with the idea of abandonment as a social fact because in both cases the owner's default causes direct damage in the first case, or indirect damage in the second case to the status of other individuals (e.g. local neighbours, users). To avoid falling back to the previous "efficientist" vision, it is important that both nuisances and negative externalities are detected in relation to specific positions and not to the city or a community in general (Han 2019; De Franco 2021). The point of reference cannot be, for example, the damage to an alleged image of the city, or urban spaces in generic terms (all land and/or buildings; see Fig. 4.4), but must be to the specific and concrete damage suffered by specific individuals.

Moreover, it should be noted that policies guided by nuisance and economic reasons could also have positive results for environmental and equity issues in general, while the opposite is not possible. The former two favour, in an *indirect* way, a more intense use/transformation of existing assets. The latter two, typically aim to *directly* promote densification making the market supply inelastic and not necessarily of higher quality. Consider for instance the option of raising taxes on empty/vacant dwellings or sealing certain areas from building activities. The unintended consequences are: in the first case, rising the costs of holding and production of urban land and services (hitting mostly physical persons); and in the second case,

---

Cohen 2001; Galster 2001, 2019; Hillier et al. 2003; Basset et al. 2006; Samsa 2008; Schilling 2009; Silverman et al. 2013; Morckel 2014; Dewar et al 2015; Grossman and Haase 2016; Molloy 2016; Han 2017; Sampson et al. 2017; Newman et al. 2018; Foster and Newell 2019; vom Hofe et al. 2019; Jeon and Kim 2020; Wegman 2020).

[21] For the Italian case, a good example may be the so-called *cedolare secca*. This norm allows the landlord to pay a fixed percentage on the rental income (20%). Originally, it was introduced in 2011 exclusively for residential rentals, but in 2018 it was also extended for other functions (Budget Law 2018, L. 205/2017). According to the *Confedilizia* surveys (2019, on Istat data 2017), the rate tax halved tax evasion by 50.45% while reducing the gap between tax and contribution revenues. This tool is seen in a positive way to relieve the property tax of owners and to stimulate the use of spaces otherwise unused by renting at advantageous prices. In this regard, Confedilizia (2019) proposed to extend the *cedolare secca* also to all commercial activities. Presumably, this instrument will be of crucial importance to relieve certain urban markets negatively affected by forced closures during the lockdown periods.

[22] This does not mean that no idea of efficiency is suitable or applicable to cities/urban processes, but simply, underlines the inadequacy of the idea of "allocative efficiency" when applied to the city (Cordato 2007; Moroni 2010, 2015a, 2019, 2020; Ikeda 2010, 2017). For a different perspective, see the idea of "adaptive efficiency" that North (1990) proposes in place of the idea of allocative efficiency. Compare also with Stone (1985).

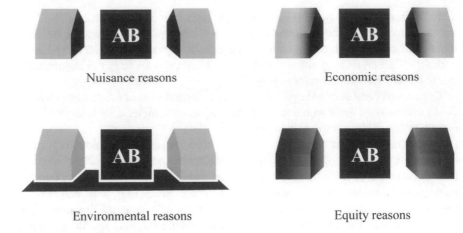

Fig. 4.4 Relational conceptualization of the variety of reasons

stiffening urban markets in favour of existing owners to the detriment of future agents.[23]

4.2.4 Policies

As regards *policies*, it can be observed that there is not (only) one main resolutive policy, but we must imagine a plurality of options. This too depends in large part on recognizing that abandonment is a social fact and that the factors influencing it are also multiple.

It is interesting to show the connection between the type of policies and factors they can affect (Fig. 4.5). For instance, *infrastructural policies* can positively affect functional factors such as those connected with location (increasing accessibility to certain spaces or areas).[24] *Collaborative policies* might affect psychological and economic factors: for example, enticing unwilling owners to get involved through vertical partnerships with the public or horizontal partnerships with other private individuals.[25]

---

[23] As some authors point out, these consequences depend on the fact that policy makers and academicians do not necessarily have in mind that they operate (and live) in an open-ended and evolutionary system. In this regard, see the critical comments by Brueckner, (2000), McLaughlin (2012), Hirokawa and Gonzalez (2010) and Henderson (2015). See for comparison Cheshire and Sheppard (2002, p. 254).

[24] For instance, increasing transportation amenities or creating new roads (Mallach 2012). Note, however, that not all types of interventions have to be "hard" or linked to a public project or plan. See on this Sampson et al. (2017), Foster and Newell (2019), Humpris and Rauws (2020).

[25] Consider for instance *common interest developments*; and other types of strategies, more or less formal, oriented for instance to have "eyes on the street" (e.g. Aiyer et al. 2015) rather than "vigilantes".

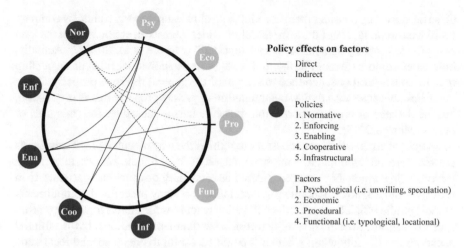

Policy effects on factors

——— Direct
------- Indirect

● Policies
   1. Normative
   2. Enforcing
   3. Enabling
   4. Cooperative
   5. Infrastructural

● Factors
   1. Psychological (i.e. unwilling, speculation)
   2. Economic
   3. Procedural
   4. Functional (i.e. typological, locational)

Fig. 4.5 Direct and indirect effects of types of policies on abandonment factors

Enabling policies may have an effect on psychological, economic and typological factors: economic incentives for restructuring can stimulate the entrepreneurship of usually passive agents and constitute a stimulus in a situation of general crisis as well as helping to recover some more difficult types of buildings.[26] *Enforcing policies* can directly counter the psychological factors related to speculation, although they must be considered as extreme measures.[27] *Normative policies* may obviously have an effect on all factors in a more or less direct way; the results can be broader and/or more radical depending on the different planning approaches adopted.[28]

In the case of Milan, as in many other Italian cities, the normative and enforcing policies play a primary role and are widely debated, while all the others were less employed and discussed. An interesting point in the local debate on normative and enforcing policies concerned the meaning and role of private property.

Enforcing policies that *compel* owners to comply with their pragmatic duties (that is, the obligation and eventual enjoinment to take action in order to restore abandoned buildings to a safe state) have been the subject of much debate in Italy after some local measures in this regard.[29] Typically, those who oppose these kinds of measures

---

[26] In the last decade, building redevelopment in Italy represents 37% of investments in the whole construction sector (ANCE 2019, pp. 10–13). Interventions on "extraordinary maintenance" have exceeded, since 2009, the ones on new buildings (annual growth rate: +0.5%).

[27] For a general criticism of the reductionist conceptions of "human choices" or "psychological events" by physicalist theories (e.g. materialism, behaviouralism, functionalism) see Jacquette (2002, pp. 233 ff.).

[28] Certain strategies focused on "urban codes" may suggest new approaches to the problem (Alfasi 2018; Hakim 2016; Moroni 2011). See also Ross and Portugali (2018). For a critical discussion on the Italian case, see Micelli and Pellegrini (2017, 2019).

[29] In a survey by Cittalia and ANCI (2009), 25% of municipal orders issued in Italy were used to coerce redevelopment (to contrast urban degradation) and provide security. Ordinances are virtually aimed at protecting the "physical integrity" of the population. However, Confedilizia (2019)

do so because they consider them as a violation of private property rights; by contrast, those who are in favour of these measures consider them as an application of the idea that private property also has a "social function" (Moroni et al. 2020).[30] Actually, both sides could be wrong. In the first case, because individual titles of ownership may be constrained or sanctioned to safeguard the general rights of property; in the latter case, because it is not the social function of private property that is in question, but the defence of *every* property from possible harms caused by the (mis)uses of others (Moroni 2018).

In light of all this, it seems necessary to rethink the way in which pragmatic duties are *set*. Here, an important role might be played by local building codes, after a long period during which they were considered as secondary instruments (particularly in Italy: Moroni 2015b).[31] Another problem is the too many procedural commitments of local public officials which reduce time and resources for more systematic on-the-spot inspections. Specifically, on detecting abandonment situations, two additional questions are the following: whether it could be useful to create special local code enforcement divisions; and whether it can be useful to more formally involve citizens and associations to map abandoned buildings and areas. These steps should be part of a general policy approach that clearly entails an unequivocal zero-tolerance approach (Moroni et al. 2020).

As regards enabling and collaborative policies, the Italian local governments have been less proactive in *stimulating* and/or *motivating* urban agents in the fulfilment of their responsibilities. Infrastructural policies, that can *support* the redevelopment of abandoned sites, should also be taken more seriously.

It could be of interest to recall here a case that occurred in Milan. On the basis of the recent building code and land use plan (Comune di Milano 2016, 2019c) and the mapping of single abandoned structures, Milan's planning councillor invited the owner (i.e. ENPAM) of certain abandoned buildings to intervene (by recovering or demolishing them). In an open letter, the adviser of ENPAM, while open to desirable forms of cooperation with the public administration to recover the area in question, underscores that the abandonment of certain buildings is probably due also to the bad state of the structures' surroundings, comprising public spaces and connecting infrastructures.[32]

---

raises several doubts on the use of ordinances as "ordinary" tools, stressing how these cause the prevarication of a sphere of competences between private and public agents.

[30] As stated in art. 42 of the Italian Constitution. Similar formulations can be found in other countries. On this, see Davy (2020).

[31] Since 2009 many Italian cities have started to reform their local building codes (with a renewal rate of 40% in 2010 and 80% by 2013, concerning around 1000 contexts for a total of 21 million inhabitants). According to Legambiente and CRESME (2013), this trend has been influenced by European directives and certificates that have promoted (and requested) a gradual shift from prescriptive to performance-related rules for the application of energy/environmental standards in building works. The hope is that abandonment issues will inspire part of these revisions, as initially attempted in the case of Milan (see Appendix A).

[32] See https://www.monitorimmobiliare.it/colliers-lettera-aperta-all-assessore-maran-urbanistica-milano_201710241339 (Accessed June 2021).

To conclude, it could be observed that the main purpose of the policies should be to *restart or reactivate the pragmatic duties of the private parties*. This again testifies to the importance of understanding abandonment as a social fact.

In this regard, a general approach to be more effective in abandonment problems should: (i) avoid definitory confusion between types of assets (e.g. empty vs. abandoned buildings), that are common mistakes also in official records (e.g. censuses); (ii) consider the scale and entity of truly problematic situations to be tackled (e.g. risks for human safety and security), in order to identify pragmatic duties of individuals (i.e. owners) and justify interventions to grant satisfactory levels for local living (e.g. on nuisances and externalities); (iii) adjust local information as an opportunity for an intensive quality check of official data (e.g. land registry, cadastres, zoning), to favour the reinstallation of the abandoned property in local market circuits (i.e. diminishing transactions costs); (iv) develop control systems for "early warnings" or "early solutions" based on owners' responsibilities and/or users' agency (e.g. local residents, passers-by) to lay out conditions for management practices (Foster and Newell 2019).

References

Aalbers MB (2016) The financialization of housing: a political economy approach. Routledge, New York

Accordino J, Johnson GT (2000) Addressing the vacant and abandoned property problem. J Urban Aff 22(3):301–315. https://doi.org/10.1111/0735-2166.00058

Aiyer SM, Zimmerman MA, Morrel-Samuels S, Reischl TM (2015) From broken windows to busy streets: a community empowerment perspective. Health Educ Behav 42(2):137–147. https://doi.org/10.1177/1090198114558590

Alfasi N (2018) The coding turn in urban planning: could it remedy the essential drawbacks of planning? Plan Theory 17(3):375–395

ANCE (2019) Osservatorio congiunturale sull'industria delle costruzioni [Economic Observatory on the Construction Industry]. Available at: https://www.ance.it. Accessed 15 November 2019

Atkinson R (2019) Necrotecture: lifeless dwellings and London's super-rich. Int J Urban Reg Res 43(1):2–13

Basset EM, Schweitzer J, Panken S (2006) Understanding housing abandonment and owner decision-making in Flint, Michigan: an exploratory analysis. Lincoln Inst Land Policy. https://doi.org/10.1017/CBO9781107415324.004

Brueckner JK (2000) Urban sprawl: diagnosis and remedies. Int Reg Sci Rev 23(2):160–171

Cheshire P, Sheppard S (2002) The welfare economics of land use planning. J Urban Econ 52(2):242–269

Cho SH, Kim SG, Lambert DM, Roberts RK (2013) Impact of a two-rate property tax on residential densities. Am J Agr Econ 95(3):685–704

Cittalia ANCI (2009) Oltre le ordinanze i sindaci e la sicurezza urbana. Cittalia Fondazione Anci ricerche. Available at: https://www.cittalia.it. Accessed 14 Jan 2018

Cohen JR (2001) Abandoned housing: exploring lessons from Baltimore. Hous Policy Debate 12(3):415–448. https://doi.org/10.1080/10511482.2001.9521413

Comune di Milano (2016) Regolamento edilizio [Building regulations]. Available at: http://www.comune.milano.it. Accessed 27 March 2020

Comune di Milano (2019a) Documento di piano [Plan document]. Available at: http://www.com une.milano.it. Accessed 18 July 2020

Comune di Milano (2019b) Piano delle regole [Rules plan]. Available at: http://www.comune.mil ano.it. Accessed 18 July 2020

Comune di Milano (2019c) Piano dei servizi [Service plan]. Available at: http://www.comune.mil ano.it. Accessed 18 July 2020

Confedilizia (2015) Dossier tassazione immobili [Property tax dossier]. Available at: https://www. confedilizia.it. Accessed 16 May 2018

Confedilizia (2019) Confedilizia notizie [multiple issues]

Cordato RE (2007) Efficiency and externalities in an Open-Ended Universe. The Ludwig von Mises Institute, Auburn

Davy B (2020) Dehumanized housing'and the ideology of property as a social function. Plan Theory 19(1):38–58

De Franco A (2021) Abandonment as an "Urban" Problem? Critical Implications and Challenges for Urban Studies. In: Bevilacqua C, Calabrò F, Della Spina L (eds) New Metropolitan Perspectives. NMP 2020. Smart Innovation, Systems and Technologies, vol 178. Springer, Cham. https://doi. org/10.1007/978-3-030-48279-4_82

Dewar M, Seymour E, Druță O (2015) Disinvesting in the city: the role of tax foreclosure in Detroit. Urban Aff Rev 51(5):587–615

Edson CL (1972) Housing abandonment—the problem and a proposed solution. Real Property, Probate Trust J 7(2):382–390

Foster A, Newell JP (2019) Detroit's lines of desire: footpaths and vacant land in the Motor City. Landsc Urban Plan 189(April):260–273. https://doi.org/10.1016/j.landurbplan.2019.04.009

Galster G (2001) On the nature of neighbourhood. Urban Studies 38(12):2111–2124

Galster G (2019) Why shrinking cities are not mirror images of growing cities: a research agenda of six testable propositions. Urban Aff Rev 55(1):355–372. https://doi.org/10.1177/107808741 7720543

Gentili M, Hoekstra J (2018) Houses without people and people without houses: a cultural and institutional exploration of an Italian paradox. Hous Stud 34(3):425–447. https://doi.org/10.1080/ 02673037.2018.1447093

Greenberg MR, Popper FJ, West BM (1990) The TOADS a new American urban epidemic. Urban Aff Q 25(3):435–454. 0803973233

Grossmann K, Haase A (2016) Neighborhood change beyond clear storylines: What can assemblage and complexity theories contribute to understandings of seemingly paradoxical neighborhood development? Urban Geogr 37(5):727–747. https://doi.org/10.1080/02723638.2015.1113807

Gu D, Newman G, Kim JH, Park Y, Lee J (2019) Neighborhood decline and mixed land uses: mitigating housing abandonment in shrinking cities. Land Use Policy 83(February):505–511. https://doi.org/10.1016/j.landusepol.2019.02.033

Hakim B (2016) Mediterranean urbanism. Springer, Heidelberg

Han HS (2017) Neighborhood characteristics and resistance to the impacts of housing abandonment. J Urban Aff 39(6):833–856

Han HS (2019) Exploring threshold effects in the impact of housing abandonment on nearby property values. Urban Aff Rev 55(3):772–799. https://doi.org/10.1177/1078087417720303

Henderson SR (2015) State intervention in vacant residential properties: an evaluation of empty dwelling management orders in England. Eviron Plann C Gov Policy 33(1):61–82. https://doi. org/10.1068/c12215

Hillier AE, Culhane DP, Smith TE, Tomlin CD (2003) Predicting housing abandonment with the Philadelphia neighborhood information system. J Urban Aff 25(1):91–106. https://doi.org/10. 1111/1467-9906.00007

Hirokawa KH, Gonzalez I (2010) Regulating vacant property. The Urban Lawyer 42(3):627–637. https://www.jstor.org/stable/27895816

Humphris I, Rauws W (2020) Edgelands of practice: post-industrial landscapes and the conditions of informal spatial appropriation. Landscape Res (online ahead of print)

Ikeda S (2010) The mirage of the efficient city. In: Goldsmith SA, Elizabeth L (eds) What we see. New Island Press, Dublin, pp 24–33

Ikeda S (2017) A city cannot be a work of art. Cosmos Taxis 4(2/3):79–86. https://cosmosandtaxis. org/ct-423/

Jacquette D (2002) Ontology. Acumen, Chesham

Jennings CR (1999) Socioeconomic characteristics and their relationship to fire incidence: a review of the literature. Fire Technol 35(1):7–34. https://doi.org/10.1023/A:1015330931387

Jeon Y, Kim S (2020) Housing abandonment in shrinking cities of East Asia: case study in Incheon, South Korea. Urban Studies 57(8):1749–1767. https://doi.org/10.1177/0042098019852024

Keenan P, Lowe S, Spencer S (1999) Housing abandonment in inner cities-the politics of low demand for housing. Hous Stud 14(5):703–716. https://doi.org/10.1080/02673039982687

Kirzner IM (1997) How markets work: disequilibrium, entrepreneurship and discovery. Inst Econ Aff, London

Kraut DT (1999) Hanging out the no vacancy sign: eliminating the blight of vacant buildings from urban areas. New York University Law Rev 74(4):1139–1177

Kroes P (2003) Screwdriver philosophy: Searle's analysis of technical functions. Techné: Res Philos Technol 6(3):131–140

Legambiente, CRESME (2013) ON-RE osservatorio nazionale regolamenti edilizi per il risparmio energetico. Available at: https://www.legambiente.it. Accessed 12 Sept 2019

Lieb RC, Merel RA, Perlin AS, Sadoff MB (1974) Abandonment of residential property in an urban context. DePaul Law Rev 23(3):1186–1224

Loddo OG (2016) Assegnare funzioni: Agentività e normatività. Sistemi Intelligenti 28(2–3):293–318

Long H, Li Y, Liu Y, Woods M, Zou J (2011) Accelerated restructuring in rural China fueled by 'increasing vs. decreasing balance' land-use policy for dealing with hollowed villages. Land Use Policy 29:11–22. https://doi.org/10.1016/j.landusepol.2011.04.003

Lynch K (1990) Wasting away. Random House, New York

Mallach A (2012) Depopulation, market collapse and property abandonment: surplus land and buildings in legacy cities. In: Mallach A (ed) Rebuilding America's legacy cities: strategies for cities losing population. American Assembly, New York, pp 85–110

Marshall S (2012) Planning, design and the complexity of cities. In: Portugali J, Meyer H, Stolk E, Tan E (eds) Complexity theories of cities have come of age: an overview with implications to urban planning and design. Springer, Heidelberg, pp 191–205

McLaughlin RB (2012) Land use regulation: Where have we been, where are we going? Cities 29:S50–S55

MEF (Ministero dell'Economia e delle Finanze) (2019) Gli immobili in Italia. Ricchezza, reddito e fiscalità immobiliare. Available at: https://www.agenziaentrate.gov.it. Accessed 21 Jan 2020

Merrett S, Smith R (1986) Stock and flow in the analysis of vacant residential property. Town Plann Rev 57(1):51–67

Micelli E, Pellegrini P (2017) Wasting heritage. The slow abandonment of the Italian Historic Centers. J Cult Heritage. https://doi.org/10.1016/j.culher.2017.11.011

Micelli E, Pellegrini P (2019) Paradoxes of the Italian historic centres between underutilisation and planning policies for sustainability. Sustainability 11(9):2614. https://doi.org/10.3390/su11092614

Molloy R (2016) Long-term vacant housing in the United States. Reg Sci Urban Econ 59:118–129

Morckel V (2014) Predicting abandoned housing: Does the operational definition of abandonment matter? Community Dev 45(2):121–133. https://doi.org/10.1080/15575330.2014.892019

Morgan DJ (1980) Residential housing abandonment in the United States: the effects on those who remain. Environ Plan A 12(12):1343–1356. https://doi.org/10.1068/a121343

Moroni S (2010) An evolutionary theory of institutions and a dynamic approach to reform. Plan Theory 9(4):275–297. https://doi.org/10.1177/1473095210368778

Moroni S (2011) Land-use planning and the question of unintended consequences. In Andersson DE (ed) The spatial market process. Advances in Austrian economics. Emerald, Bingley, pp 265–288. https://doi.org/10.1108/S1529-2134(2012)0000016013

Moroni S (2015a) Complexity and the inherent limits of explanation and prediction. Plan Theory 14(3):248–267. https://doi.org/10.1177/1473095214521104

Moroni S (2015b) Libertà e innovazione nella città sostenibile [Freedom and innovation in the sustainable city]. Carocci, Roma

Moroni S (2018) Property as a human right and property as a special title. Rediscussing private ownership of land. Land Use Policy, 70(January):273–280. https://doi.org/10.1016/j.landusepol. 2017.10.037

Moroni S (2019) Constitutional and post-constitutional problems: reconsidering the issues of public interest, agonistic pluralism and private property in planning. Plan Theory 18(1):5–23. https:// doi.org/10.1177/1473095218760092

Moroni S (2020) The just city. Three background issues: institutional justice and spatial justice, social justice and distributive justice, concept of justice and conceptions of justice. Plann Theor 19(3):251–267. https://doi.org/10.1177/1473095219877670

Moroni S, Minola L (2019) Unnatural sprawl: reconsidering public responsibility for suburban development in Italy, and the desirability and possibility of changing the rules of the game. Land Use Policy 86(July):104–112. https://doi.org/10.1016/j.landusepol.2019.04.032

Moroni S, De Franco A, Bellè BM (2020) Unused private and public buildings: re-discussing merely empty and truly abandoned situations, with particular reference to the case of Italy and the city of Milan. J Urban Aff 42(8):1299–1320. https://doi.org/10.1080/07352166.2020.1792310

Németh J, Langhorst J (2014) Rethinking urban transformation: temporary uses for vacant land. Cities 40:143–150. https://doi.org/10.1016/j.cities.2013.04.007

Newman G, Park Y, Lee RJ (2018) Vacant urban areas: causes and interconnected factors. Cities 72:421–429

North D (1990) Institutions, institutional change and economic performance. Cambridge University Press, Cambridge

Olivadese R, Remøy H, Berizzi C, Hobma F (2017) Reuse into housing: Italian and Dutch regulatory effects. Prop Manag 35(2):165–180. https://doi.org/10.1108/PM-10-2015-0054

OPPAL (2019) L'efficienza dei processi concessori [The efficiency of planning bureaucracy]. Available at: http://www.gestitec.polimi.it. Accessed 17 Feb 2020

Peñalver EM (2010) The illusory right to abandon. Mich Law Rev 109(2):191–219

Power A, Mumford K (1999) The slow death of great cities? Urban abandonment or urban renaissance. York Publishing Services Ltd., York

Raslanas S, Zavadskas EK, Kaklauskas A (2010) Land value tax in the context of sustainable urban development and assessment. Int J Strateg Prop Manag 14(1):73–86. https://doi.org/10.3846/ ijspm.2010.06

Rauws W (2017) Embracing uncertainty without abandoning planning: exploring an adaptive planning approach for guiding urban transformations. DisP Plann Rev 53(1):32–45

Resta G (2018) I rifiuti come beni in senso giuridico. Rivista Critica Del Diritto Privato 36(2):249–268

Romano S (2002) Studi sulla derelizione nel diritto romano. Rivista Di Diritto Romano 2:99–164

Ross GM, Portugali J (2018) Urban regulatory focus: a new concept linking city size to human behaviour. Royal Soc Open Sci 5(5):1–11. https://doi.org/10.1098/rsos.171478

Sampson N, Nassauer J, Schulz A, Hurd K, Dorman C, Ligon K (2017) Landscape care of urban vacant properties and implications for health and safety: lessons from photovoice. Health Place 46(May):219–228. https://doi.org/10.1016/j.healthplace.2017.05.017

Samsa MJ (2008) Reclaiming abandoned properties: using public nuisance suites and land banks to pursue economic redevelopment. Cleveland State Law Rev 56(1):189–232

Schilling J (2009) Code enforcement and community stabilization: the forgotten first responders to vacant and foreclosed homes. Albany Gov Law Rev 2:101–162

Searle J (1995) The construction of social reality. The Free Press. La costruzione della realtà sociale. Einaudi, Torino, New York, p 2006

Silverman RM, Yin L, Patterson KL (2013) Dawn of the dead city: an exploratory analysis of vacant addresses in Buffalo, NY 2008–2010. J Urban Aff 35(2):131–152. https://doi.org/10.1111/j.1467-9906.2012.00627.x

Smith ME (2020) Definitions and comparisons in urban archaeology. J Urban Archaeol 1:15–30

Sternlieb G, Burchell RW, Hughes JW, James FJ (1974) Housing abandonment in the urban core. J Am Plann Assoc 40(5):321–332. https://doi.org/10.1080/01944367408977488

Stone CN (1985) Efficiency versus social learning: a reconsideration of the implementation process. Rev Policy Res 4(3):484–496. https://doi.org/10.1111/j.1541-1338.1985.tb00248.x

Tieffenbach E (2010) Searle and Menger on money. Philos Soc Sci 40(2):191–212

Vincent J (2012) Neighborhood revitalization and new life: a land value taxation approach. Am J Econ Sociol 71(4):1073–1094. https://doi.org/10.1111/j.1536-7150.2012.00849.x

vom Hofe R, Parent O, Grabill M (2019) What to do with vacant and abandoned residential structures? The effects of teardowns and rehabilitations on nearby properties. J Reg Sci 59(2):228–249. https://doi.org/10.1111/jors.12413

Wang K, Immergluck D (2019) Housing vacancy and urban growth: explaining changes in long-term vacancy after the US foreclosure crisis. J Housing Built Environ 34(2):511–532. https://doi.org/10.1007/s10901-018-9636-z

Wegmann J (2020) Residences without residents: assessing the geography of ghost dwellings in big US cities. J Urban Aff 42(8):1103–1124

Wilson D, Margulis H, Ketchum J (1994) Spatial aspects of housing abandonment in the 1990s: the Cleveland experience. Hous Stud 9:493–510

Chapter 5
Conclusions

Abstract These concluding remarks are divided into three parts. The first part is an overview of the findings. The second part proposes a new outlook on "abandonment": a dynamic/evolutionary approach suggesting that abandonment is a potential state of any type of (private) property. In this view, abandonment is not a purely and intrinsically negative phenomenon, but something that can also be an opportunity for critical imagination and creative action. The third part focuses on the limitations of the study and possible future research lines.

Keywords Dynamic processes · Evolutionary approach · Urban systems · Change

5.1 Overview

Chapter 1 suggests the idea that abandonment should be understood as a social fact and not as a brute fact (Searle 1995). Thus, in this work the "abandoned" state of buildings is considered as not directly related to certain physical variables; rather, it entirely depends on *human evaluations*. Crucial information in this regard is how institutional frameworks—sets of rules of conduct—influence individual behaviour and actions through time.

In Chap. 2, the theoretical framework is presented. Three contributions to the debate are: the distinction between brute and social facts and that between agentive and non-agentive functions; these two combined with the idea of pragmatic duty provide a new understanding of the abandonment problem and relative implications. In light of this, in Chap. 3 an analytical schema—based on *parameters*, *factors*, *reasons*, *policies*—has been built to interpret the phenomenon of abandonment and possible ways of intervening. The neo-institutional approach helps to highlight how the problem of abandonment is articulated with respect to various aspects influencing urban transformations, e.g. property rights, formal constraints, arguments behind policy decisions, intervention strategies and implementations.

In Chap. 4, the analytical schema was used to propose some ideas and critical hypotheses on the descriptive and evaluative aspects of the phenomenon of abandonment. It seems impossible to define a single factor able to explain the abandoned state of all the buildings in a city, nor of the majority of them. Actually, we found a *multiplicity* of factors influencing abandonment; a fact that suggests employing a *plurality* of policies to cope with it. To do so, a more adaptive approach to land-use regulations may support capacity and coalition building with private actors (Rauws 2017) that are, also in the case of abandonment, among "the forgotten first responders" (Schilling 2009). As regards the possible reasons for negatively judging abandonment in urban contexts, only some of them seem to provide coherent and valuable normative directions (i.e. nuisance and economic reasons), since they imply a more "situational" approach to the problem.

All the previous observations testify to the importance of revisiting the issues at stake and attempts to bring back a critical debate on elements often entrusted to ideology.

5.2   Reframing Abandonment

This book explores the problem of abandonment beyond the traditional focus on residential properties (Edson 1972; Lieb et al. 1974; Morgan 1980; White 1986; Wilson et al. 1994; Keenan et al. 1999; Cohen 2001) while underlining its difference in comparison to the issue of building "vacancy". Some would argue that "long-term vacancy" and "abandonment" are the same problem (Glock and Häußermann 2004; Dewar and Weber 2012; Molloy 2016), although the latter expresses a social fact while the former is more a functional aspect. In reality, the phenomenon is more ubiquitous as it potentially concerns *all private agents and properties* (Mallach 2006; Basset et al. 2006; Schilling 2009; Bogataj et al. 2016)[1] and it is *inherent to the way urban systems operate*. The conclusion is that it is not possible to eradicate it; the point is rather what to do about it (vom Hofe et al. 2019).

In light of what has been explored, we can say that abandonment is not an unusual or unnatural phenomenon but part of social urban processes. Furthermore, it is not purely or intrinsically negative but something that can also be an opportunity for critical imagination and creative action.

In speaking about obsolete buildings, already Jacobs (1961, p. 187) underlined that: "Cities need old buildings so badly it is probably impossible for vigorous streets and districts to grow without them. By old buildings I mean … a good lot of plain, ordinary, low-value old buildings, including some rundown old buildings". She was perfectly aware of the dynamic nature of certain phenomena, as clearly stressed in this other passage: "Over the years there is … constantly a mixture of buildings of many ages and types. This is, of course, a dynamic process, with what was once new in the mixture eventually becoming what is old in the mixture" (Jacobs 1961, p. 189). She goes on warning that "The more infertile the simplified territory becomes for economic enterprises, the still fewer the users and the still more infertile the territory.

---

[1] This is particularly evident in Italy, where most of the properties are owned by physical persons (either for residential and non-residential typologies). See ANCE (2012, 2019) and MEF (2019).

A kind of unbuilding, or running-down process is set in motion. … Whatever it may be specifically, the role of the dead place as a geographic obstacle to the general land has overcome its role as a contributor of users to the general land" (Jacobs 1961, pp. 259–263). The fact itself that we question "change" implies that we are truly concerned about a living world and not a "dead" one (Popper 1974, p. 147).[2] Paraphrasing Searle's words (1995, p. 15), if we value death and extinction above all, then we would say that the function of abandonment is to accelerate the death of urban systems. Instead, when conceiving abandonment as a "social fact", it can be regarded as evidence of where and how agentive functions take place (e.g. relatively to property rights, titles, obligations, responsibilities, permissions, authorisations, requirements; see Searle 2006, p. 59).[3]

Reformulating a famous image of Taleb (2007), one might say that not each and every abandoned building is an "ugly duckling".

Therefore, the right attitude towards abandonment is not "indignant amazement" (as Bennett and Dickinson 2015 also stress) but the predisposition of the public agent to see an opportunity to change conditions (through a broad spectrum of policies).

If one considers abandoned buildings as part of a wholesome and dynamic urban context, the emergence of abandonment problems should not be seen as antithetical or antagonist to an idea of a "well-functioning city". Clearly, non-abandoned buildings are what makes urban systems work but this is different to saying that all buildings in cities should be used and functional at full capacity.

In this regard, three points can be underlined: (i) cities cannot work without empty structures; (ii) policies should not fix or determine evolutionary trajectories; (iii) it is more desirable to consider urban reality as an open-ended system where agents find their ways through the physical and social realms, continuously adjusting their activities to solve practical problems.

The neo-institutional approach helps to reconstruct how certain phenomena "express different—and often contrasting—layers of changing reality" (Salet 2018, p. 1 ff.). Because in urban contexts "change is the only constant" (De Roo 2012), seeing abandonment as a social fact underlines that the phenomenon is not so exceptional but constitutive of how urban contexts work.

This study, therefore, proposes a sort of dynamic/evolutionary approach in explaining the abandonment phenomenon in urban contexts, suggesting *it is a potential state of any type of private property.*

In this perspective, we can observe that many policies are often a posteriori/reactive, with respect to the phenomenon, while we need interventions a priori/anticipatory (Buitelaar et al. 2021).

---

[2] It may be interesting to recall here the words of Seneca (*Epistula ad Lucilium XII*) who, on discussing with the guardian of his childhood home, discovers that he has aged with it. In his words (translated by the author): "I owe this to my suburban villa, (that is) the fact that wherever I turned my old age was evident to me. Let us embrace it and love it; it is full of pleasure, if you know how to exploit it".

[3] This conceptual approach may be extended and applied to generative practices underlying urban processes in general. On this, see also Smith (2020b, p. 24).

5.3   Limitations and Further Research Directions

A limitation of this work is that it is mainly conceptual, considering a single emblematic case (i.e. Milan). Although the use of a single case study is usual and accepted in research work (Flyvbjerg 2006),[4] the specific topic addressed suggests testing the findings also through other comparative case studies.[5]

Future research may develop more in depth how abandonment contributes, in a positive way, to incentivise entrepreneurship of various agents: public, private but also non-profit entities.[6]

Other lines of work may develop and integrate the theoretical framework and analytical schema presented here and apply it to other similar issues. This study focused mainly on the abandonment of private buildings, but a similar problem also arises for public buildings that deserve as much attention (Moroni et al. 2020a, 2020b). The focus of attention was mainly on the owners, although the issue also affects other subjects (think, for instance, of the possible role of tenants). Here, we mainly dealt with urban areas, but abandonment in rural areas would also deserve attention.[7]

The wish is to have been able to contribute to refining definitions and analytical categories to understand more deeply the autonomy and peculiarities of "social facts", underlying the complexity of the reality we experience and inhabit every day.

References

Aiyer SM, Zimmerman MA, Morrel-Samuels S, Reischl TM (2015) From broken windows to busy streets: a community empowerment perspective. Health Educ Behav 42(2):137–147. https://doi. org/10.1177/1090198114558590
ANCE (2012) Osservatorio congiunturale sull'industria delle costruzioni [Economic observatory on the construction industry]. Available at: http://www.ance.it. Accessed 28 Oct 2018

---

[4] As Flyvbjerg (2006) demonstrates it is not necessarily a mistake to generalize starting from a single case study. As he writes: "One can often generalize on the basis of a single case and the case study may be central to scientific development via generalization as supplement or alternative to other methods. But formal generalization is overvalued as a source of scientific development, whereas 'the force of example' is underestimated" (Flyvbjerg 2006, p. 12).

[5] Even if a single case study may be not enough, the categories here introduced may be used for dynamic comparative studies (Smith 2020a). Moreover, the Milan case (see Appendix A) itself requires to be followed up over time, to specifically assess the effects of the policies just launched.

[6] Thus, other open challenges regard *informal* processes and organizations, grasping opportunities from spontaneous activities. Certain studies in these directions remark how the "serendipitous" still plays a relevant role in structuring physical and social processes (Humpris and Rauws 2020, p. 2). For instance, to combat local distress by "mere presence" (Aiyer et al. 2015), or improve accessibility design starting from "desire lines" (Foster and Newell 2019). See also the "ruling-without-rules" perspective: Lorini and Moroni (2020); Lorini and Moroni (2021).

[7] For extra-urban situations (problems, strategies, etc.) with reference to Italy, see, e.g. Morena et al. (2017, 2019).

ANCE (2019) Osservatorio congiunturale sull'industria delle costruzioni [Economic observatory on the construction industry]. Available at: https://www.ance.it. Accessed 15 Nov 2019

Basset EM, Schweitzer J, Panken S (2006) Understanding housing abandonment and owner decision-making in Flint, Michigan: an exploratory analysis. Lincoln Inst Land Policy. https://doi.org/10.1017/CBO9781107415324.004

Bennett L, Dickinson J (2015) Forcing the empties back to work? Ruinphobia and the bluntness of law and policy. Paper presented at the transience and permanence in Urban development international research workshop, University of Sheffield, Town and Regional Planning Dept, Sheffield, 14–15 Jan 2015

Bogataj D, McDonnell DR, Bogataj M (2016) Management, financing and taxation of housing stock in the shrinking cities of aging societies. Int J Prod Econ 181:2–13. https://doi.org/10.1016/j.ijpe.2016.08.017

Buitelaar E, Moroni S, De Franco A (2021) Building obsolescence in the evolving city. Reframing property vacancy and abandonment in the light of urban dynamics and complexity. Cities, 108 (online ahead of print)

Cohen JR (2001) Abandoned housing: exploring lessons from Baltimore. Hous Policy Debate 12(3):415–448. https://doi.org/10.1080/10511482.2001.9521413

de Roo G (2012) Spatial planning, complexity and a world 'out of equilibrium'—outline of a non-linear approach to planning. In: de Roo G, Hillier J, Van Wezemael J (eds) Complexity and spatial planning: systems, assemblages and simulations. Ashgate Publishing, Farnham, pp 129–165

Dewar M, Weber MD (2012) City abandonment. In: Dewar M, Weber MD (eds) The Oxford handbook of urban planning. Oxford University Press, Oxford, pp 563–586

Edson CL (1972) Housing abandonment—the problem and a proposed solution. Real Property, Probate Trust J 7(2):382–390

Flyvbjerg B (2006) Five misunderstandings about case-study research. Qual Inq 12(2):219–245

Foster A, Newell JP (2019) Detroit's lines of desire: footpaths and vacant land in the motor city. Landsc Urban Plan 189(April):260–273. https://doi.org/10.1016/j.landurbplan.2019.04.009

Glock B, Häußermann H (2004) New trends in urban development and public policy in eastern Germany: dealing with the vacant housing problem at the local level. Int J Urban Reg Res 28(December):919–929

Humphris I, Rauws W (2020) Edgelands of practice: post-industrial landscapes and the conditions of informal spatial appropriation. Landscape Res (online ahead of print)

Jacobs J (1961) The death and life of great American cities. Vintage, New York

Keenan P, Lowe S, Spencer S (1999) Housing abandonment in inner cities-the politics of low demand for housing. Hous Stud 14(5):703–716. https://doi.org/10.1080/02673039982687

Lieb RC, Merel RA, Perlin AS, Sadoff MB (1974) Abandonment of residential property in an urban context. DePaul Law Rev 23(3):1186–1224

Lorini G, Moroni S (2020) Ruling without rules: Not only nudges. Regulation beyond normativity. Global Jurist 1 (online ahead of print)

Lorini G, Moroni S (2021) Rule-free regulation: exploring regulation 'without rules' and apart from 'deontic categories'. J Theor Soc Behav (online ahead of print).

Mallach A (2006) Bringing buildings back: from abandoned properties to community assets. Rutgers University Press, New Brunswick (NJ)

MEF (Ministero dell'Economia e delle Finanze) (2019) Gli immobili in Italia. Ricchezza, reddito e fiscalità immobiliare. Available at: https://www.agenziaentrate.gov.it. Accessed 21 Jan 2020

Molloy R (2016) Long-term vacant housing in the United States. Reg Sci Urban Econ 59:118–129

Morena M, Truppi T, Del Gatto ML (2017) Sustainable tourism and development: the model of the Albergo Diffuso. J Place Manag Dev 10(5):447–460. https://doi.org/10.1108/JPMD-08-2016-0057

Morena M, Bischetti GB, Del Gatto ML, Gornati A (2019) Innovative management of rural buildings. J Cult Heritage Manage Sustain Dev 9(1):43–61. https://doi.org/10.1108/JCHMSD-09-2017-0065

Morgan DJ (1980) Residential housing abandonment in the United States: the effects on those who remain. Environ Plan A 12(12):1343–1356. https://doi.org/10.1068/a121343

Moroni S, De Franco A, Bellè BM (2020a) Vacant buildings. Distinguishing heterogeneous cases: public items versus private items; empty properties versus abandoned properties. In: Lami IM (ed) Abandoned buildings in contemporary cities: smart conditions for actions. Springer, Cham, pp 9–18

Moroni S, De Franco A, Bellè BM (2020b) Unused private and public buildings: Re-discussing merely empty and truly abandoned situations, with particular reference to the case of Italy and the city of Milan. J Urban Affairs 42(8):1299–1320. https://doi.org/10.1080/07352166.2020.179 2310

Popper K (1974) Unended quest: an intellectual autobiography. Open Court, Chicago

Rauws W (2017) Embracing uncertainty without abandoning planning: exploring an adaptive planning approach for guiding urban transformations. DisP- Plann Rev 53(1):32–45

Salet W (2018) Public norms and aspirations: the turn to institutions in action. Routledge, London

Schilling J (2009) Code enforcement and community stabilization: the forgotten first responders to vacant and foreclosed homes. Albany Gov Law Rev 2:101–162

Searle J (1995) The construction of social reality. The Free Press. La costruzione della realtà sociale. Einaudi, Torino, New York, p 2006

Searle J (2006) Social ontology: Some basic principles. Anthropological Theory 6(1):12–29

Smith ME (2020a) The comparative analysis of early cities and urban deposits. J Urban Archaeol 2:197–205

Smith ME (2020b) Definitions and comparisons in urban archaeology. J Urban Archaeology 1:15–30

Taleb NN (2007) The black swan: the impact of the highly improbable. Random House, New York

vom Hofe R, Parent O, Grabill M (2019) What to do with vacant and abandoned residential structures? The effects of teardowns and rehabilitations on nearby properties. J Reg Sci 59(2):228–249. https://doi.org/10.1111/jors.12413

White MJ (1986) Property taxes and urban housing abandonment. J Urban Econ 20(3):312–330. https://doi.org/10.1016/0094-1190(86)90022-7

Wilson D, Margulis H, Ketchum J (1994) Spatial aspects of housing abandonment in the 1990s: the Cleveland experience. Hous Stud 9:493–510

Appendix A
Case Study

The problem of abandoned buildings in Italian urban contexts is fuelling policy debates at various institutional levels.[1] Public discussions, however, are not always able to identify and separate truly problematic states of built assets from totally legitimate states of affairs (Moroni et al. 2020a). This problem is complicated by the lack of official figures on the magnitude of the abandonment problem in local contexts. Also in Italy, various quantifications of the number of abandoned buildings in the country often result over- or under-estimated. This has critical effects in policy making because public discussions expand into too many ambiguous directions (as also noted by Gentili and Hoekstra 2018).

As already underscored in the Introduction (Chap. 1), this section focuses on one main case study: the city of Milan. The analysis is based on official data found in official documents and reports from national and international agencies,[2] interviews with relevant actors, an analysis of local normative records[3] and national regulations and on-site investigations.[4]

---

[1] As already underscored in the introduction, this appendix is partially based on ideas and materials developed by the author in Moroni et al. (2020b) and De Franco (2021).

[2] For instance Assolombarda (2019), Camera di Commercio (2017, 2018, 2019), Confcommercio (2015), OPPAL (2019).

[3] See for instance Comune di Milano (2016, 2017, 2018a, b, 2019a, b, c).

[4] Data gathering and computation have been implemented through geographical information systems (i.e. GIS analysis) and satellite imagery (i.e. Google maps, Google earth).

Table A.1 Definitions and estimations on the problem in Italy

Total buildings	Deteriorated	Abandoned	Derelict
Private stock (whole structures)	"Rundown, in ruins or under construction"	"Abandoned properties"	"Ruined properties" (i.e. *unità collabenti*)
14,500,000 (ISTAT 2014)	750,000 (ISTAT 2014)	2,000,000 (CESCAT: Assoedilizia)	474,000 (Agenzia delle Entrate 2017b)

This section deals briefly with the general abandonment problem in Italy and more in depth with the issue in the city of Milan.

An Outline of the Abandonment Problem in Italy

In Italy, as in other countries, different terms are used to denote the phenomenon of abandonment of private properties. These terms are not always transparent or unambiguous.[5] A rough estimate of the number of abandoned buildings in Italy can be done by breaking down official information from census and fiscal data (Table A.1).

A first source is ISTAT (2014) according to which 750,000 buildings are "rundown, in ruins or under construction": this is equal to 5.2% of the total stock of private buildings in the country (around 14.5 million buildings[6]). To have the exact amount of abandoned buildings those under construction—not specified by the report—should be removed.

A second source is the Agenzia delle Entrate (2017b), stating that 474,000 units in Italy are derelict (i.e. in a severe state of abandonment).[7]

---

[5] In the case of Italy, it is not possible to limit the "abandonment" problem to one specific or prevalent type of building or context. The set is wide and diverse (Legambiente and CRESME 2013; Marini and Santangelo 2013; Micelli and Pellegrini 2017; Adobati and Garda 2018; Micelli and Pellegrini 2019) and often the way of conceiving the phenomena can be misleading. In many cases, the abandonment of buildings is discussed as a pure technical problem (relegating it to the "decommissioning" of industrial buildings) and/or as an a priori collective problem (aligning it to issues of vacant properties).

[6] About 84% of them are residential buildings, corresponding to a total gross surface area of around 4 billion m^2 (Agenzia delle Entrate 2017a).

[7] *Unimpresa* signals similar data. See https://www.unimpresa.it/edilizia-unimpresa-mezzo-milione-di-immobili-in-dissesto/15184 (Accessed June 2021). According to this report, the provinces with the highest number of buildings in a severe state of abandonment are Frosinone, Cosenza, Cuneo, Benevento, Foggia, Aosta, Siracusa, Piacenza, Verbanio Cusio Ossola, Vibo Valentia.

A third source is research by *Assoedilizia* (more precisely, by *Centro Studi Casa Ambiente e Territorio*)[8] talking about 2 million abandoned properties in Italy (this amount is largely cited also in parliamentary bills[9]).

Since *Agenzia delle Entrate* and ISTAT data refer only to ruined/rundown/derelict buildings, while the *Centro Casa Ambiente e Territorio* adopts a broader definition, we may hypothesise that abandoned buildings in Italy—according to the definition here assumed—are around one million (about 7% of the total stock).[10]

The Case of Milan: An Overview

The case of Milan has been chosen because local authorities announced and discussed new strategies to tackle the problem of the abandonment of buildings. The resulting policies have been the subject of local and national debate.

General Data and Figures

The city of Milan is the regional capital of Lombardy (in northern Italy). By 2019 the population was around 1,400,000 (in an area of more than 18,000 hectares). Population forecasts signal an overall increase of the population of 5.6% up to 2030 (Comune di Milano 2019a).[11] The city is one of the most dynamic in the country,[12] with the

---

[8] See http://www.assoedilizia.com/contenuto_exp.asp?t=comunicati&id=375 (Accessed January 2019).

[9] See for instance the Italian Draft Law no. 4054 of 2016: http://documenti.camera.it (Accessed February 2019).

[10] Considering exclusively merely empty buildings, we can note the following. According to the last available ISTAT (2014) census 77.3% (out of over 31 million dwellings) are occupied by at least one resident, while the 22.7% are empty dwellings or dwellings used as other than primary residences (that is, second-homes or dwellings used by non-resident persons who dwell there, e.g. for study or work). Gentili and Hoekstra (2018) hypothesize that in Italy the latter percentage can be articulated as follows: 40% truly empty dwellings, 50% holiday homes, 10% dwellings occupied by non-residents (or rented on the black market). Compare with the percentages suggested for the German situation by Glock and Häußermann (2004). In Italy, this would indicate over than 2,500,000 empty dwellings. This amount is similar to the one suggested in a report by *The Guardian*. See https://www.theguardian.com/society/2014/feb/23/europe-11m-empty-propertiesen ough-house-homeless-continent-twice (Accessed February 2019).

[11] During 2008–2019, the population grew by 6.7%; in the same period, the foreign population increased by 47% (Comune di Milano 2019a).

[12] In the 2008–2017 period, the number of young residents increased by 21.7% (in the 19–24 age class). In 2019, the number of students enrolled at the local universities was around 200,000. In the same year, the service sector employed 70% of the workers (Comune di Milano 2019a). The number of innovative start-ups is continuously growing (Camera di Commercio 2018). In light of all this, Milan can be considered a non-shrinking city: in moderate growth and economically active.

strongest international exposure,[13] and it contributes significantly to the national GDP (Assolombarda 2019). The companies here operating in 2019 were around 300,000 (37% of the entire region); about 41,000 of which are in the construction sector.[14]

In Milan, the real estate market is trying to recover from the negative effects of economic crisis (which have been less severe than in other cities in Italy). The prices of new houses in the first half of 2019 registered an average increase of +1% (with peaks of up to 4%).[15] Nonetheless, the problem of abandoned buildings also arises in a city like Milan.

In the public debate, abandoned private buildings are considered a problem because of street and area security (e.g. personal safety) and their use as sites of illegal activities.[16] Others further highlighted their detrimental effects on nearby properties (i.e. reducing their market value). Mattarocci (2017, pp. 41–46) demonstrates that abandoned spaces in Milan exhibit the typical "negative externality" effect evident in decreasing values for selling or renting nearby properties within a certain buffer radius (see Fig. A.1).[17]

Abandoned Private Buildings in a Non-shrinking City

About 170 abandoned private buildings (of a certain size) were mapped. A first exploratory investigation was done through the *Atlante dell'Abbandono*, a project developed by the *L'ABB* laboratory (Università Statale of Milan) and by the *Centro Studi PIM*.[18] Subsequently, an official mapping was produced (Comune di Milano 2019b). The municipality identified abandoned buildings through inspections on the spot and cross-referencing tax data (Fig. A.2). Around 35% of them are industrial structures. More than 80% of the buildings are located in peripheral districts while the others are situated in central or semi-central districts.

---

[13] In 2015, Milan hosted the Expo event, which attracted around 21.5 million visitors (Confcommercio 2015). It benefited the local economy: for an estimation of the overall economic benefits (e.g. on the real-estate market), see Dell'Acqua et al. (2016).

[14] See https://www.milomb.camcom.it/i-numeri-delle-imprese (Accessed December 2019). See also Camera di Commercio (2019) on the entire situation.

[15] See   https://www.milomb.camcom.it/rilevazione-semestrale-dei-prezzi-degli-immobili (Accessed October 2019).

[16] For a general discussion see the public hearings at https://www.camera.it/application/xmanager/ projects/leg17/attachments/upload_file_commissione_periferia/pdfs/000/000/009/2._Audizioni_ in_prefettura.pdf (Accessed November 2018).

[17] Mattarocci's analysis (2017) of the Milan urban area shows that prices decrease on average by €80/m2 within 1 km and €40/m2 within 2 km. The impact is greater for residential properties (i.e. around −€100/m2 and −€50/m2 respectively in 1 or 2 km). While industrial properties have the lowest decreases (i.e. less than €20/m2), other types of functions (e.g. retail and office buildings) suffer more on rental values in the immediate surroundings.

[18] See http://www.pim.mi.it/atlante-abbandono (Accessed April 2019).

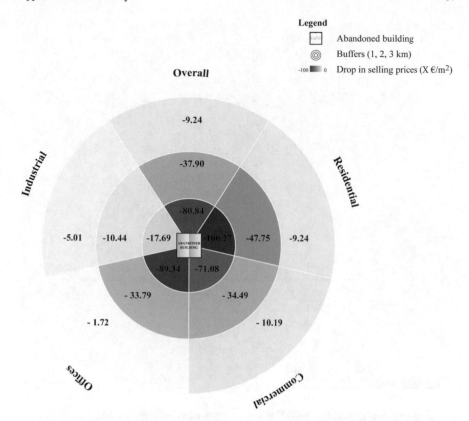

Fig. A.1 Diagrammatic representation of the economic effects of abandonment in Milan showing price developments of real estate sectors (€/m²) within certain distances. Based on Mattarocci (2017)

It is interesting to show the relationship between abandonment and the urban tissues in Milan. Let us consider three typical cases: (i) abandonment in central/historical areas (high accessibility, mix and quality of the built environment, Fig. A.3); (ii) abandonment in semi-central areas (areas with good accessibility but mainly residential, Fig. A.4); (iii) abandonment in peripheral areas (scarce accessibility, low mix and low quality of the built environment, e.g. mainly industrial or old nuclei, Fig. A.5).

Local Policies

"Abandoned" buildings are defined in the building code (Comune di Milano 2016) as those unoccupied and unutilized buildings without maintenance for more than five years. If the owner of one of these buildings does not restore the property to a

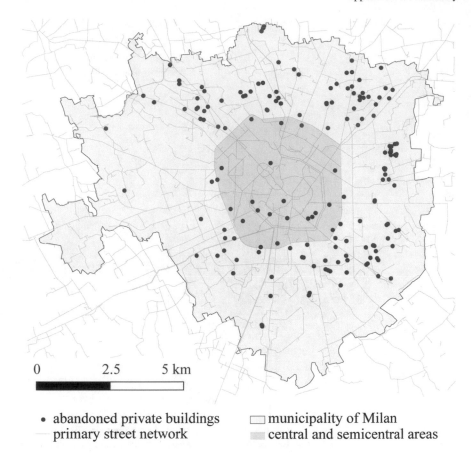

• abandoned private buildings   □ municipality of Milan
  primary street network   ▨ central and semicentral areas

Fig. A.2 Abandoned buildings as identified by Comune di Milano (2019b)

safer state—once received a notification in this regard by the municipality—the code envisions a direct public intervention and the attribution of the costs to the owner. If this is not possible, the code envisages the confiscation of the property and the allocation of it to some kind of public use (Comune di Milano 2016, art. 12). In the city, various ordinances for restoring the properties have been issued since the adoption of the building code.[19]

The issue of abandoned buildings has been definitively framed by the local land use plan (approved in October 2019).[20] Local population and stakeholders took

---

[19] Based on the interview with public officials, most of the owners replied to the first warrants (issued at the beginning of 2018). Basic interventions as bordering the property were already in place, but after the solicitation of the municipality most of the owners declared their interest in intervening on the property and some of them started to demolish their properties (see Appendix B, Fig. B.3, for details). This, for municipal technicians was considered as a positive feedback: it meant that most owners were at least existing even if not always vigilant on the property.

[20] With 26 votes for and 12 against in the approval session of the municipal council.

Central areas   1:30.000

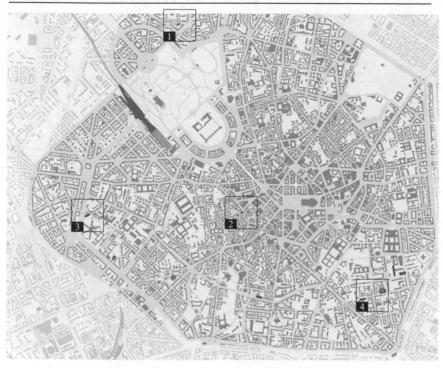

Selected areas   1:10.000

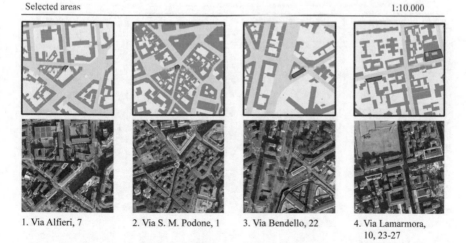

1. Via Alfieri, 7   2. Via S. M. Podone, 1   3. Via Bendello, 22   4. Via Lamarmora,
                                                                                           10, 23-27

Fig. A.3 Abandoned buildings in central areas of Milan (selection)

Semi-central areas   1:30.000

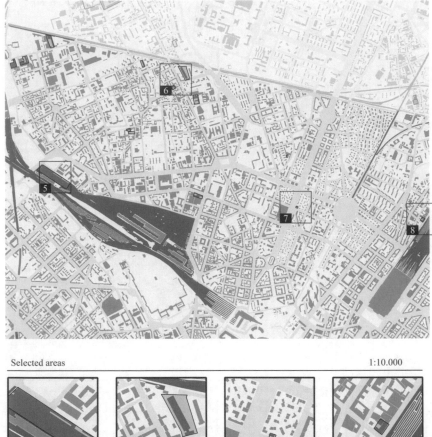

Selected areas   1:10.000

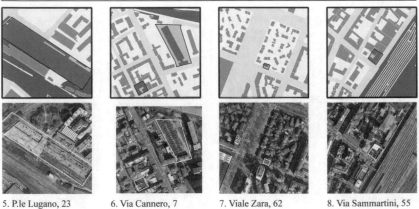

5. P.le Lugano, 23   6. Via Cannero, 7   7. Viale Zara, 62   8. Via Sammartini, 55
                         Via Conte Verde, 10

Fig. A.4 Abandoned buildings in semi-central areas of Milan (selection)

Periphery   1:30.000

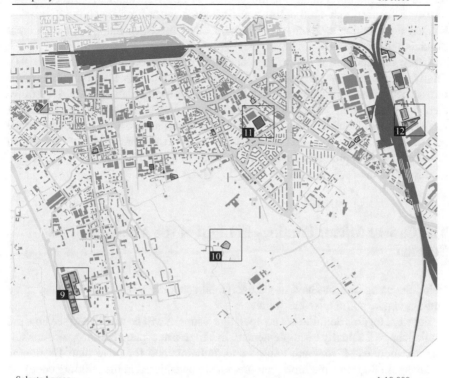

Selected areas   1:10.000

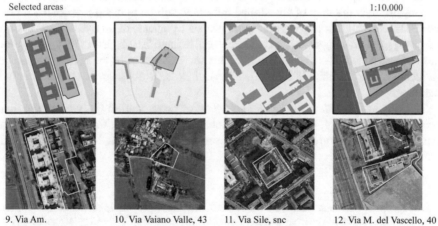

9. Via Am.   10. Via Vaiano Valle, 43   11. Via Sile, snc   12. Via M. del Vascello, 40
Vincenzo, 15/19
Via Antegnati, 7/15

Fig. A.5 Abandoned buildings in peripheral areas of Milan (selection)

part in the preparation stages of the plan. This occurred thanks to the collection of formal requests and 4400 online questionnaires, workshops involving 350 stakeholders representing 172 organisations and other meetings mainly for municipal officials and technicians.

Urban regeneration is one of the plan focal points (Comune di Milano 2019a). Firstly, the plan safeguards around 3 million m^2 of the municipal territory for agriculture (in comparison with the previous plan, it reduces the buildable area by 4%, equivalent to 1.7 million m^2). Secondly, the plan systematically deals with the problem of recovering abandoned buildings in the city (see in particular art. 11 of the regulatory part of the plan—the so-called "*Piano delle regole*": Comune di Milano 2019b). In this regard, a map precisely locates the private abandoned buildings identified (see map R.10, "*Carta del consumo di suolo*") (Figs. A.6 and A.7).

The Case of Milan: Findings in Light of the Analytical Schema

Using the categories outlined in the analytical schema presented in Chap. 3, some findings can be summarised as follows.

For what regards definitions and operative issues, it will be decisive to see how the regulations will actually be implemented. In Milan's new plan, buildings are considered "abandoned" if they are unused and unmaintained for more than 12 months (reducing, in this way, the time span previously considered in the building code). In this regard, how the map of abandoned buildings will be updated is a crucial issue. For instance: will this happen on a yearly basis by means of an executive decision (i.e. a "*determina dirigenziale*"; see Comune di Milano 2019b, art. 11)? A geo-portal has been activated: it is unclear whether this serves to monitor the whole process or to identify new abandoned buildings. Nevertheless, it does not have a normative meaning.[21]

Looking at the *factors of the abandonment* of private buildings, we can see that economic factors (e.g. the impacts of global economic crisis) remain relevant in Italy and in other southern countries in Europe (Bogataj et al. 2016; Knieling and Othengrafen 2016).[22]

---

[21] Last update by the municipality in October 2019. See https://geoportale.comune.milano.it/Map ViewerApplication/Map/App?config=/MapViewerApplication/Map/Config4App/405andid=ags (Accessed February 2021).

[22] Real house prices in Italy increased by 55% in the period 1998–2007 and decreased by 20% at the end of 2012 (Baldini and Poggio 2014, p. 325). During the last economic crisis, the construction sector experienced a strong contraction in the number of workers and companies. According to ANCE (2019), between 2008 and 2010, around 27,000 companies left the market; moreover 218,000 jobs were lost. This has contributed to a stagnation of the value of urban dwellings in the country, differently to what happened in other European cities (Agenzia delle Entrate 2019a, p. 3; Eurostat 2018, p. 44).

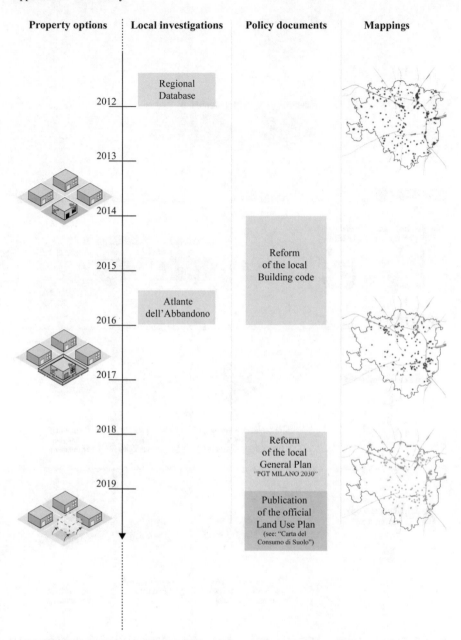

Fig. A.6 Timeline of the case of Milan

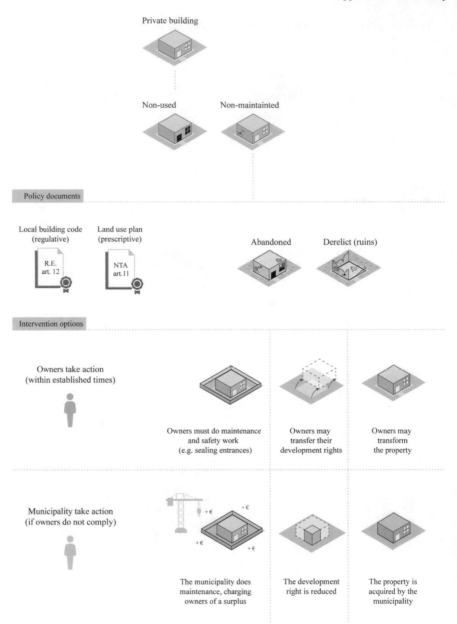

Fig. A.7 Procedural process of the normative and enforcing policies to cope with abandonment in Milan. Based on Comune di Milano (2016 and 2019b)

Table A.2 Subnational economic rankings across 13 major cities in Italy (World Bank 2020)

Location	Starting a business	Dealing with construction permits	Getting electricity	Registering property	Enforcing contracts
Cagliari	9	1	8	11	8
Padua	3	2	11	12	6
Bologna	6	3	1	2	3
Florence	5	4	4	5	13
Ancona	1	5	12	4	7
Rome	13	6	3	1	5
Turin	4	7	2	9	1
Genoa	6	8	9	3	4
Palermo	6	9	13	6	10
Reggio Calabria	9	10	5	10	9
Naples	9	11	6	7	12
Bari	9	12	7	12	11
Milan	1	13	10	7	2

It is interesting to observe how *psychological factors* (e.g. personal attitudes of the owners) contribute in various ways to the problem: both lacks of entrepreneurship and speculative strategies are present.

Procedural factors (e.g. bureaucracy) strongly affect the abandonment problem in Italy, for instance when considering the time required for approval of local projects.[23] In 2020 the city of Milan ranked in the last position for construction permits in comparison with other major cities (see Table A.2).[24]

In Italy, the average time to approve a (private) implementation plan was around seven months in 2018, with peaks of over twenty-five months in, for example, Milan

---

[23] In a report by the World Bank (2019) on the situation of various countries, Italy ranked 104th in terms of costs and complication of procedures and the amount of time required for obtaining a building permit (an average of 190 days). See www.doingbusiness.org (Accessed February 2021).

[24] This specific point is largely shared by the interviewees, but from different viewpoints. A business operator acting as real estate developer highlighted that long approval times in Italy are well known to locals but also to foreign companies. This causes an "annihilation" of healthy domestic and international competition. On the same subject, actors from the construction sectors pointed out that long approval times are mostly due to the high fragmentation and redundancy of legislations across the institutional levels. In his view, a hierarchical breakdown is necessary, although the local level (in his view, the most crucial one), is the "sleepiest" or "numbest". This, in his view, prevents the activation and informing of upper-level institutions (e.g. regional) of the necessary reforms on land-use and building regulations (e.g. fewer and more open functional classifications to decompress approval processes and overcome "obsolete" technical standards more quickly). Finally, from the viewpoint of an architect, not only long approvals but also too detailed requirements (e.g. public works bids) actually "dumb down" the creative process and consequently also the physical and functional outcomes of interventions, both in private and public areas.

(OPPAL 2019).[25] Other problems related to regulatory uncertainty concerning technical and construction activities in Italy, that, as some enquiries show, have changed almost once every 23 days.[26] While these changes have been regarded positively by some (e.g. to align with EU directives and stimulate resource exchanges), their effectiveness can be improved. For instance, the many incentives for building upgrades to be truly advantageous should be accredited in a relatively short amount of time (e.g. five years instead of ten) and be more integrated.[27]

In the same vein, the president of *Confedilizia* (i.e. the Italian Confederation of Property Owners) stressed that the increase of the municipal property tax (i.e. IMU, introduced in 2011) directly exacerbated the problem of abandoned buildings nationwide.[28]

Looking at the *functional factors* (e.g. typologies and distribution of abandoned buildings) in Milan we see that the vast majority of abandoned buildings are non-residential; their reuse or demolition is difficult because of their size and of certain typologies, such as hangars, large commercial estates or logistic buildings. Obviously, the chances for their transformation will depend on the willingness to solve problems of poor accessibility and unattractiveness of their sites.

Public/official arguments against abandonment problems lean more on *nuisance reasons* (safety and security, e.g. in the building code) and *environmental reasons* (e.g. the general plan); while *economic* and *equity reasons* emerge more in the general local discussions. The public discussion resumed with this matter in consideration.

The types of *policies for coping with the abandonment* problem in Milan, seem to favour normative and enforcing measures, while partnership, incentive and enabling policies are marginally tackled.

---

[25] Considering the local level, the average time for obtaining a building permit in Italy was 87 days. Some cities (e.g. Ancona, Firenze, Lucca, Milano, Modena, Padova, Pisa, Rimini, Torino) have peaks of over 100 days (OPPAL 2019).

[26] Studies are based on changes in the *Testo Unico dell'edilizia* (Dpr 380/2001) without considering regional laws and national rules on bidding disciplines. See http://amp.ilsole24ore.com/pagina/AEr KgimC (Accessed November 8 2019).

[27] An example in Italy is the "Restructuring bonus (*bonus ristrutturazioni*)" (introduced by the Legislative Decree no. 83 of 2012, art. 11). According to it, works for building upgrades receive a tax break of 50%. Another example is the "Façade bonus (*bonus facciate*)" (introduced by Law no. 160 of 2019) implying a tax break of 90%.

[28] See https://www.confedilizia.it/nel-2015-65-ruderi-rispetto-al-2011/ (Accessed January 2018). Note that, even in ancient times, the dereliction of *funds* was exacerbated by tributary/fiscal obligations (Romano 2002, pp. 120–121). In Italy, those who declare their buildings as uninhabitable or derelict, after the check of administrative technicians, can have consistent tax discounts (from 50% up to 100%). The latest enquiry on cadastral data shows that "derelict" properties in Italy rose by 97% between 2011 and 2018 (Agenzia delle Entrate 2019b). Compared with 2011, property taxes in Italy almost tripled in 2015; they grew from 0.6% in 2010 to 1.5% in 2012 as a percentage of the national GDP (Agenzia delle Entrate 2015).

Normative and *enforcing* policies have been the subject of much debate in Milan and in Italy in general.[29]

As regards *enabling policies*, we observe that the change in use is always possible in principle.[30] In the case of a change from industrial to residential use, the volume of the building can be completely reused (Comune di Milano 2019a). Moreover, under a certain floor area (i.e. 250 m^2), the change from an industrial to a residential use does not involve particular charges (Comune di Milano 2019c). The so-called "neighbourhood shops [*esercizi di vicinato*]" have further incentives, introduced with the aim of recovering ground floors (Comune di Milano 2019a, 2019b).[31] Considering the problems at stake, this enabling approach is still moderate.

As regards *collaborative policies*, even in mature and dynamic contexts such as the city of Milan, these initiatives generate various sources of friction (d'Ovidio and Cossu 2017; Tricarico and Pacchi 2018; Pacchi 2020). Probably, some of the buildings will eventually be included in ongoing strategic programmes[32] or more targeted public works.

Moreover, for *infrastructural policies*, it is possible to imagine that some of the ongoing works on new infrastructures (e.g. the Metroline 4) will eventually motivate owners to take action.[33]

---

[29] The possibility of expropriating neglected properties in the city was assumed by right-wing politicians as an instrument of "applied socialism" and "legalized expropriation"; see https://mil anesi.corriere.it/2014/01/08/edifici-abbandonati-se-la-proprieta-non-ci-pensa-deve-intervenire-il-comune/ (Accessed May 2021); while the left wing parties (at that time in charge) underlined that the intention was to stimulate "the real defaulters [that] are the large real estate property funds, which leave properties to crumble"; see https://st.ilsole24ore.com/art/notizie/2014-06-09/nel-mir ino-comune-milano-160-immobili-abbandonati-uso-pubblico-se-proprietari-non-interverranno--172111.shtml?uuid=ABaqmLPBandrefresh_ce=1 (Accessed May 2021). This was also seen as an opportunity to trigger concrete interventions for the safety of the population, urban regeneration and temporary uses. Similar debates started in other contexts (e.g. Rome and Naples), gaining the attention also of some members of the Constitutional Court. See http://www.salviamoilpa esaggio.it/blog/2014/06/immobili-abbandonati-e-funzione-sociale-dopo-napoli-e-lora-di-milano/ (Accessed May 2021).

[30] In the absence of building works (as stated in art. 8 of the plan: Comune di Milano 2019b).

[31] This last measure was welcomed by *Confcommercio* (i.e. the local traders' association). See the claims by the secretary general: https://www.confcommerciomilano.it/it/news/comunicati_stampa/2018/.content/cs/Confcommercio_Milano_Pgt_negozi_di_vicinato (Accessed January 2020).

[32] One of the buildings mapped is awaiting inclusion in the redevelopment project *Scalo Farini*. After its decommissioning in 2000 a new project had been approved only by 2017, after many negotiations and clauses (e.g. clearance works, reduction of the gross floor area and admitting exclusively commercial and tertiary activities). See https://milano.corriere.it/19_gennaio_17/milano-ruspe-all-hotel-lugano-l-ecomostro-disperati-rinasce-palazzo-ex-poste-abbandonato-ea0 400da-1a27-11e9-b5e1-e4bd7fd19101.shtml (Accessed October 2019). See also Comune di Milano (2017) and Pasqui (2018).

[33] Note that, around the 40% of the properties mapped by the municipality are near to these areas (Fig. A.2).

Final Remarks

To conclude, one of the central issues of the debate in Milan was the normative powers in the hands of the public authorities. If the demolition of the abandoned property does not take place, both the local plan and the building code state that the municipal administration has the right to directly intervene with the demolition and subsequently claim reimbursement from the private owners. The development rights of the owners who have not promptly organised for demolition are subject to a reduction at the minimum ratio established (0.35 m^2/m^2). The plan reduces the time requested to consider a building without maintenance as "abandoned" while adopting a more penalising approach.[34] Actually, considering the legal nature of the land-use plan in Italy, some additional problems arise. Doubts emerge regarding the eventuality that a local plan may be applied in this way to certain issues, such as assets property and development rights. An additional problem arises from the fact that buildings are considered as abandoned not solely if they are unused properties that are the cause of "risks for the security or health or safety of the public", but also if they are responsible for "diminished urban decorum and quality" (in art. 11); that is, a formulation open to too many interpretations.[35]

Recently, the Lombardy Region has introduced a new law (no. 18 of 2019) on urban regeneration. It adopts a partially different approach to the phenomenon and suggests quite different measures (e.g. additional volumes for recovering abandoned buildings). A problem will therefore regard how Milan's policies and this regional law will interact.[36]

The Covid-19 pandemic created difficulties for local policies to the point that certain administrative procedures (including the ordinances to restore the buildings) have been suspended from February to May 2020. It would be interesting to understand how the various aspects of the issues addressed here will develop in the post-pandemic period.

---

[34] These new regulations were criticized, in July 2019, by the president of *Assoedilizia* (that is, the oldest local property owners association). See https://assoedilizia.wordpress.com/2019/07/ (Accessed December 2019).

[35] A similar formulation was already present in the local Building Code.

[36] In this regard, a public debate between the assessor of Milan and the regional assessor of planning has started. See http://www.ansa.it/lombardia/notizie/consiglio_lombardia/2019/11/12/_28f5e25f-623b-494f-9e7d-696fae0a3b32.html (Accessed December 2019). Three recent sentences by the Regional Administrative Court (i.e. TAR) recognized that the regional law illegitimately reduces the autonomy of the local government while being in contrast with other regional laws on "soil consumption" (see the Lombardy TAR sentences no. 371, 372, 373 of the 10th February 2021). The issue has been recently brought up to the Constitutional Court. See https://blog.urbanfile.org/2021/02/11/milano-urbanistica-immobili-abbandonati-il-tar-da-ragione-al-comune/ (Accessed May 2021).

References

Agenzia delle Entrate (2015) La tassazione immobiliare: Un confronto internazionale [Real estate taxation: an international comparison]. Available at: https://www.finanze.gov. Accessed 13 Oct 2018

Agenzia delle Entrate (2017a) Gli immobili in Italia [Real estate in Italy]. Available at: http://www.mef.gov.it. Accessed 24 Maggio 2019

Agenzia delle Entrate (2017b) Statistiche catastali 2016 [Cadastral statistics 2016]. Available at: https://wwwt.agenziaentrate.gov.it. Accessed 2 June 2019

Agenzia delle Entrate (2019a) Rapporto immobiliare 2019: Settore residenziale [Real estate report 2019. Residential sector]. Available at: https://www.agenziaentrate.gov.it. Accessed 27 Jan 2020

Agenzia delle Entrate (2019b) Statistiche catastali 2018: Catasto edilizio urbano [Land registry report 2019]. Available at: www.agenziaentrate.gov.it. Accessed 27 Jan 2020

Adobati F, Garda E, (eds) (2018) Biografie sospese. Un'esplorazione dei luoghi densamente disabitati della Lombardia. Mimesis/Kosmos, Sesto San Giovanni

Assolombarda (2019) Booklet: Italy, Lombardy and Milan. Available at: https://www.assolombarda.it. Accessed 28 July 2020

Baldini M, Poggio T (2014) The Italian housing system and the global financial crisis. J Housing Built Environ 29(2):317–334

Bogataj D, McDonnell DR, Bogataj M (2016) Management, financing and taxation of housing stock in the shrinking cities of aging societies. Int J Prod Econ 181:2–13. https://doi.org/10.1016/j.ijpe.2016.08.017

Camera di Commercio (2017) Milano, l'Europa. Città internazionali a confronto [Milan, Europe. International cities in comparison]. Available at: https://www.milomb.camcom.it. Accessed 19 June 2020

Camera di Commercio (2018) L'importanza di essere una start up [The importance of being a start-up]. Available at: https://www.milomb.camcom.it. Accessed 19 June 2020

Camera di Commercio (2019) Milano produttiva [Productive Milan]. Available at: https://www.milomb.camcom.it. Accessed 19 June 2020

Comune di Milano (2016) Regolamento edilizio [Building regulations]. Available at: http://www.comune.milano.it. Accessed 27 March 2020

Comune di Milano (2017) Documento di visione strategica: Scali ferroviari [Strategic vision document. Railway yards]. Available at: http://www.comune.milano.it. Accessed 21 March 2020

Comune di Milano (2018a) Documento unico di programmazione 2019–2012 [Single programming document 2019–2012]. Available at: http://www.comune.milano.it. Accessed 4 April 2019

Comune di Milano (2018b) Ricognizione sullo stato di attuazione dei programmi [Reconnaissance on the state of implementation of the programs]. Available at: http://www.comune.milano.it. Accessed 6 May 2019

Comune di Milano (2019a) Documento di piano [Plan document]. Available at: http://www.com une.milano.it. Accessed 18 July 2020

Comune di Milano (2019b) Piano delle regole [Rules plan]. Available at: http://www.comune.mil ano.it. Accessed 18 July 2020

Comune di Milano (2019c) Piano dei servizi [Service plan]. Available at: http://www.comune.mil ano.it. Accessed 18 July 2020

Confcommercio (2015) Report attività per Expo Milano 2015 [Report for Expo Milano 2015]. Available at: https://www.confcommerciomilano.it. Accessed 30 Sept 2020

Confedilizia (2019) Confedilizia notizie [multiple issues]

Confedilizia (2015) Dossier tassazione immobili [Property tax dossier]. Available at: https://www. confedilizia.it. Accessed 16 May 2018

d'Ovidio M, Cossu A (2017) Culture is reclaiming the creative city: the case of Macao in Milan, Italy. City, Culture and Society 8:7–12

De Franco A (2021) Addressing the problem of private abandoned buildings in Italy. A neo-institutional approach to multiple causes and potential solutions. In: Bisello A, Vettorato D, Haarstad H, Borsboom-van Beurden J (eds) Smart and sustainable planning for cities and regions: results of SSPCR 2019, vol 2. Springer, Berlin, pp 235–247

Dell'Acqua A, Morri G, Quaini E, Airoldi A (2016) L'indotto di Expo 2015. Un'analisi di impatto economico al termine dell'evento [The related activities of Expo 2015. An analysis of the economic impact at end of the event]. Available at: https://www.milomb.camcom.it. Accessed 5 Feb 2019

Garda E (2018) Negli spazi vuoti della metropoli: Esperienze di riuso collettivo tra temporaneità e permanenze. Geography Notebooks 1(2):97–110

Gentili M, Hoekstra J (2018) Houses without people and people without houses: a cultural and institutional exploration of an Italian paradox. Housing Studies 34(3):425–447. https://doi.org/10.1080/02673037.2018.1447093

Glock B, Häußermann H (2004) New trends in urban development and public policy in eastern Germany: dealing with the vacant housing problem at the local level. Int J Urban Regional Res 28(December):919–929

ISTAT (2014) Edifici e abitazioni [Buildings and homes]. https://www.istat.it. Accessed 21 Feb 2019

Knieling J, Othengrafen F (eds) (2016) Cities in crisis. Socio-spatial impacts of the economic crisis in southern European cities. Routledge, London

Legambiente, CRESME (2013) ON-RE Osservatorio Nazionale Regolamenti Edilizi per il risparmio energetico. Available at: https://www.legambiente.it. Accessed 12 Sept 2019

Mattarocci G (2017) Interventi di rigenerazione urbana e valore degli immobili: Il caso di Milano [Urban regeneration interventions and property value: the case of Milan]. In Cerasoli M, Mattarocci G (eds) Rigenerazione urbana e mercato immobiliare [Urban regeneration and real estate market], Roma Tre-Press, Roma, pp 32–52

Micelli E, Pellegrini P (2017) Wasting heritage. The slow abandonment of the Italian Historic Centers. J Cult Heritage. https://doi.org/10.1016/j.culher.2017.11.011

Micelli E, Pellegrini P (2019) Paradoxes of the Italian historic centres between underutilisation and planning policies for sustainability. Sustainability 11(9):2614. https://doi.org/10.3390/su1109 2614

OPPAL (2019) L'efficienza dei processi concessori [The efficiency of planning bureaucracy]. Available at: http://www.gestitec.polimi.it. Accessed 17 Feb 2020

Pacchi C (2020) Iniziative dal basso e trasformazioni urbane. L'attivismo civico di fronte alle dinamiche di governance locale. Bruno Mondadori, Milano

Pasqui G (2018) Raccontare Milano. Politiche, progetti, immaginari [Milan. Policies, projects, imaginaries]. Angeli, Milano
Romano S (2002) Studi sulla derelizione nel diritto romano. Rivista di diritto romano (2):99–164
Tricarico L, Pacchi C (2018) Community entrepreneurship and co-production in urban development. Territorio 87:69–77
World Bank (2019) Doing business 2019. World Bank, Washington (DC). Available at: www.doingbusiness.org. Accessed 19 June 2020

Appendix B
Further Data and Figures

For additional information, see the Table B.1 and Figs. B.1, B.2, B.3.

Table B.1 Types of ownership and uses (2016 data; MEF 2019).

Types of units	Assets owned by physical persons				Assets owned by non-physical persons			
	N° units	% on the total	Rent, € (Rendita)	Rendita %	N° units	% on the total	Rent, € (Rendita)	Rendita %
Principal residence	19.509.976	34,2%	10.773.225.798	47,70%	0	0,00%	0	0,0%
Annexes	13.291.600	23,3%	1.205.179.896	5,30%	0	0,00%	0	0,0%
Owners' disposal	6.265.385	11,0%	2.148.515.618	9,50%	40.175	0,50%	37.825.256	0,3%
Rented	6.016.007	10,5%	4.172.174.648	18,50%	1.252.804	17,11%	2.908.678.106	20,4%
Free lease	1.224.965	2,1%	474.770.031	2,10%	1	0,00%	413	0,0%
Other	8.086.841	14,2%	2.773.034.149	12,30%	168.789	2,30%	411.419.477	2,9%
Unknown	602.715	1,1%	206.382.588	0,90%	3.957.108	54,00%	8.417.894.001	59,2%
Not declared	2.090.284	3,7%	820.727.825	3,60%	1.904.399	26,00%	2.452.932.955	17,2%
Total	57.087.773	100%	22.574.010.553	100%	7.323.276	100%	14.228.750.208	100%

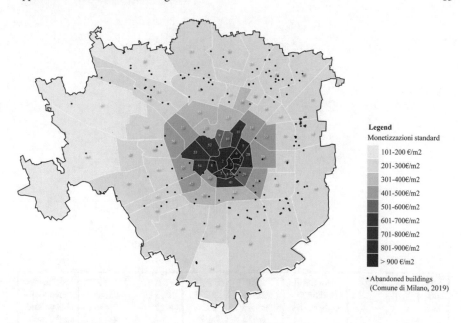

Fig. B.1 Average values of private properties in Milan's areas. Based on OMI data and *"Monetizzazioni Standard"*. See at: https://www.comune.milano.it/aree-tematiche/urbanistica-ed-edilizia/monetizzazioni (Accessed October 2018)

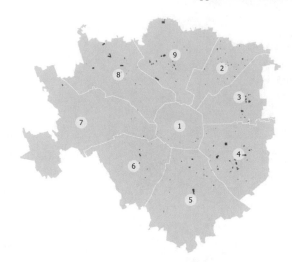

Mu-nic-ipio	Area (m 2)	Abandoned buildings (AB, units)	AB Gross Floor Area (m2)	% Gross Floor Area on the total stock of AB	AB Ground surface (m2)	% Ground surface on the total stock of AB	% Occu-pancy of AB ground surfaces on Municipio's areas
1	9.426.871,60	9	41.518,2	6%	6.731,78	1%	0,07%
2	12.628.554,98	30	55.346,38	8%	45.341,63	7%	0,36%
3	14.434.240,03	20	89.321,62	12%	66.664,18	11%	0,46%
4	20.695.477,51	30	197.131,46	27%	169.985,33	27%	0,82%
5	29.958.855,46	19	105.426,65	14%	70.816,31	11%	0,24%
6	18.336.216,76	12	13.010,32	2%	27.912,16	4%	0,15%
7	31.363.417,11	4	14.861,24	2%	10.819,66	2%	0,03%
8	23.906.482,26	19	82.312,79	11%	94.044,64	15%	0,39%
9	21.013.486,59	26	134.890,64	18%	134.890,64	22%	0,64%
Total	181.763.602,31	169	733.819,30	100%	627.206,33	100%	0,03

Fig. B.2 Final selection of the official mapping and general estimations of the sample in the different districts of the city (i.e. *municipi*). Based on Comune di Milano (2019b)

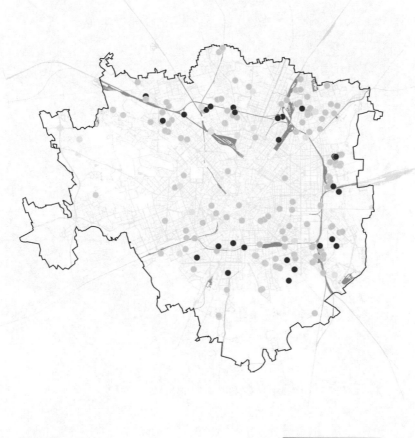

Legend

- ◦ Redeveloped
- ◦ Redeveloping (e.g. demolished)
- ● Collapsed
- ◦ Structurally stable (e.g. boarded-up, not transformed)

1 0 1 2 3 4 km

Fig. B.3 Transformation process of the abandoned buildings in Milan, based on a satellite survey (through Google Earth Imagery, November 2020). The analysis accounted for land and building transformations by observing various elements (falling roofs, construction works on the site, growing vegetation, etc.)

Index

Printed in the United States
by Baker & Taylor Publisher Services